Conformal Maps and Geometry

Advanced Textbooks in Mathematics

Print ISSN: 2059-769X
Online ISSN: 2059-7703

The *Advanced Textbooks in Mathematics* explores important topics for postgraduate students in pure and applied mathematics. Subjects covered within this textbook series cover key fields which appear on MSc, MRes, PhD and other multidisciplinary postgraduate courses which involve mathematics.

Written by senior academics and lecturers recognised for their teaching skills, these textbooks offer a precise, introductory approach to advanced mathematical theories and concepts, including probability theory, statistics and computational methods.

Published

Crowds in Equations: An Introduction to the Microscopic Modeling of Crowds
by Bertrand Maury and Sylvain Faure

The Wigner Transform
by Maurice de Gosson

Periods and Special Functions in Transcendence
by Paula B Tretkoff

Mathematics of Planet Earth: A Primer
by Jochen Bröcker, Ben Calderhead, Davoud Cheraghi, Colin Cotter, Darryl Holm, Tobias Kuna, Beatrice Pelloni, Ted Shepherd and Hilary Weller
edited by Dan Crisan

Conformal Maps and Geometry
by Dmitry Beliaev

Advanced Textbooks in Mathematics

Conformal Maps and Geometry

Dmitry Beliaev

University of Oxford

World Scientific

NEW JERSEY • LONDON • SINGAPORE • BEIJING • SHANGHAI • HONG KONG • TAIPEI • CHENNAI • TOKYO

Published by

World Scientific Publishing Europe Ltd.

57 Shelton Street, Covent Garden, London WC2H 9HE

Head office: 5 Toh Tuck Link, Singapore 596224

USA office: 27 Warren Street, Suite 401-402, Hackensack, NJ 07601

Library of Congress Cataloging-in-Publication Data

Names: Beliaev, Dmitry, author.
Title: Conformal maps and geometry / by Dmitry Beliaev (University of Oxford, UK).
Description: New Jersey : World Scientific, 2018. | Series: Advanced
textbooks in mathematics | Includes bibliographical references and index.
Identifiers: LCCN 2018040361 | ISBN 9781786346131 (hc : alk. paper)
Subjects: LCSH: Conformal mapping. | Conformal geometry. | Mappings (Mathematics)
Classification: LCC QA360 .B3354 2018 | DDC 515/.9--dc23
LC record available at https://lccn.loc.gov/2018040361

British Library Cataloguing-in-Publication Data

A catalogue record for this book is available from the British Library.

First published 2020 (hardcover)
Reprinted 2024 (in paperback edition)
ISBN 9781800616356 (pbk)

For any available supplementary material, please visit
https://www.worldscientific.com/worldscibooks/10.1142/Q0183#t=suppl

Desk Editors: V. Vishnu Mohan/Jennifer Brough/Koe Shi Ying

Typeset by Stallion Press
Email: enquiries@stallionpress.com

To my family. Without your love, encouragement
and support I would have never finished this book.

Preface

Complex analysis in general, and function theory in particular, are old and well-developed subjects. In this textbook, we concentrate on a more geometric approach that originates from the Riemann mapping theorem, which could be seen as a statement that there is a correspondence between univalent functions and planar domains. This allows us to develop a powerful theory of univalent functions with a strong geometric flavour.

The idea of this book was conceived when I started giving courses about the Schramm–Loewner evolution (SLE) and realized that there is no single complex analysis text which gives all necessary background function theory information in a simple and accessible manner. There are many excellent books on function theory, in particular books by Ahlfors, Duren, and Pommerenke which I really like but, at the same time, they might be too advanced for some students and do not cover everything that I want to explain. In particular, univalent maps in the upper half-plane, which are important in the context of the Schramm–Loewner evolution, are not treated in the classical works. Another excellent source is Lawler's book, which treats complex analysis from a probabilistic perspective.

A significant part of this book is based on the lecture notes I developed for several courses that I have taught in Oxford. It is intended to be used as the main textbook for an advanced undergraduate or a graduate course studying conformal maps on their own or building the analysis base to study SLE. Prerequisites are absolutely minimal: a basic course in complex analysis, some understanding of measures and integration (but most of the book does not require it), and basic knowledge of differential equation (for the final chapter only).

The book is organized as follows:

Introduction contains a list of basic results that might be missing from a basic complex analysis course, but will be used throughout the book. We briefly discuss these results and give references to other books that study them in greater detail.

Riemann mapping theorem gives the classical proof of the Riemann mapping theorem using the normal families approach. After this, we study the boundary behaviour of uniformizing maps and answer when conformal maps are continuous up to the boundary. Finally, we briefly discuss the connection between conformal maps and the Dirichlet boundary problem and uniformization of multiply connected domains.

Basic theory of univalent maps gives an introduction to the general theory of univalent functions. In this chapter, we prove universal bounds on functions and their derivatives (Bieberbach–Koebe theory) and discuss subordination and corresponding analytic and geometric estimates. After this, we discuss the notions of capacity and their relation to geometry. Then we study the properties of half-plane uniformizing maps. Finally, this chapter gives an introduction to coefficient problems and explains the relation between convergence of domains and functions.

Extremal length introduces the theory of conformal invariants defined in terms of a very geometric notion of extremal length. This chapter gives several applications of extremal length and explains how classical topics of harmonic measure and prime ends can be studied by the extremal length techniques.

Loewner evolution gives a modern introduction to the theory of Loewner evolution; in particular, it carefully treats both (classical) radial and chordal cases. This chapter explains how the Loewner evolution ideas can be used in classical function theory and discusses the geometric properties of hulls generated by the Loewner evolution.

The book contains a number of exercises that are an integral part of the text; solutions to all exercises can be found at the end of each chapter.

D. Beliaev
Oxford 2019

Acknowledgements

I would like to thank Lennart Carleson, Victor Havin, Peter Jones, Nikolai Makarov and Stanislav Smirnov for teaching and inspiring me. Special thanks go to Fredrik Viklund, Steffen Rohde and Ilia Binder for their valuable comments on the first draft of this book and to Michael McAuley for his help with proofreading. I would like to express my gratitude to Laurent Chaminade who suggested that I should publish this book, Mary Simpson, Jennifer Brough and the rest of the World Scientific team who made this book possible. I would also like to thank the Engineering and Physical Sciences Research Council who partially funded my research through a very generous fellowship EP/M002896/1.

Contents

Preface vii

Acknowledgements ix

1. Introduction 1

    1.1 Preliminaries 2

2. Riemann Mapping Theorem 11

    2.1 Historical Remarks 12
    2.2 Möbius Transformations and the Schwarz Lemma 13
    2.3 Normal Families 16
    2.4 Proof of the Riemann Mapping Theorem 20
    2.5 Other Normalizations 23
    2.6 Constructive Proofs 28
        2.6.1 Composition of elementary maps 28
        2.6.2 The Christoffel–Schwarz formula 29
    2.7 Boundary Correspondence 33
        2.7.1 Accessible points 34
        2.7.2 Prime ends 43
    2.8 Dirichlet Boundary Problem 47
        2.8.1 Poisson and Schwarz kernels 48
        2.8.2 Green's function 54
        2.8.3 Harmonic measure 57
    2.9 Multiply Connected Domains 59
        2.9.1 Conformal annuli 59
        2.9.2 Uniformization of multiply connected domains . . 66
    2.10 Solutions 72

3. Basic Theory of Univalent Maps 83

3.1 Classes S and Σ . . . 83
3.2 Bieberbach–Koebe Theory . . . 85
3.3 Sequences of Univalent Functions . . . 93
3.4 Subordination and the Carathéodory Class . . . 101
3.5 Capacities and Geometry . . . 104
3.6 Further Properties of Mapping-Out Functions . . . 108
3.7 Coefficient Problems . . . 112
3.7.1 Class S . . . 112
3.7.2 Class Σ . . . 116
3.7.3 Integral means . . . 117
3.8 Solutions . . . 119

4. Extremal Length and Other Conformal Invariants 129

4.1 Extremal Length . . . 130
4.1.1 Definitions and basic properties . . . 130
4.1.2 Extremal metric . . . 132
4.1.3 Composition rules . . . 134
4.2 Applications . . . 136
4.2.1 Examples and further properties . . . 136
4.2.2 Geometric application . . . 147
4.3 Harmonic Measure Revisited . . . 148
4.4 Prime Ends Revisited . . . 157
4.5 Solutions . . . 162

5. Loewner Evolution 169

5.1 Basic Loewner Evolutions for Slit Maps . . . 170
5.1.1 Radial Loewner evolution . . . 170
5.1.2 Chordal Loewner evolution . . . 179
5.1.3 Applications . . . 183
5.2 Loewner Chains . . . 188
5.3 Solving the Loewner Evolution . . . 194
5.4 When a Loewner Chain is Generated by a Curve? . . . 204
5.5 Solutions . . . 214

Bibliography 221

Index 225

Chapter 1

Introduction

All parts of analysis have a significant geometric component and this is even more so in the case of complex analysis. Unfortunately, this connection is not usually covered by standard complex analysis courses. Probably the only result that might appear in a basic course is the geometrical form of the argument principle, which relates the number of zeroes inside a contour to the winding number of its image.

The first important result that we discuss is the Riemann mapping theorem, which states that every simply connected domain is a conformal image of a standard domain. This gives a relation between univalent maps, that is, analytic one-to-one functions, and simply connected domains. In particular, we can ask how geometrical properties of domains are related to analytical properties of the corresponding univalent functions. The rest of Chapter 2 is devoted to the study of the boundary behaviour of univalent maps. In particular, we will study when these maps are continuous up to the boundary.

Chapter 3 focuses on the general properties of univalent maps in the unit disc and its complement. We prove many important *universal* results that are valid for *all* univalent functions. In particular, we obtain universal estimates on how a univalent function can behave near the boundary.

In Chapter 4, we discuss a versatile and very geometrical way of constructing conformal invariants. This is a very powerful technique which can be used to study many conformal invariants; in particular, it is very useful if one wants to *estimate* conformal invariants using geometric information or obtain some geometric information from conformal invariance.

An important example of such applications is the proof of Beurling inequalities.

Chapter 5 is devoted to a very powerful method which allows one to describe families of decreasing or increasing domains in terms of differential equations. Since its introduction in the 1930s, it was used to obtain many important results. It was a crucial tool in proving the Bieberbach conjecture and in the recent development of the Schramm–Loewner evolution (SLE) theory.

1.1. Preliminaries

It is assumed that readers are familiar with basics of complex analysis, namely, the material covered by a standard one-term complex analysis course. There are several facts that are sometimes excluded from such courses but will be used extensively in this book. We state these results to make the text as self-contained as possible.

Green's formula in complex form. The classical Green's formula states that if Ω is a domain bounded by a finite number of positively oriented curves and functions $P(x,y)$ and $Q(x,y)$ are continuously differentiable in the closure of Ω, then

$$\int_\Omega [\partial_x P(x,y) - \partial_y Q(x,y)]\,\mathrm{d}x\mathrm{d}y = \int_{\partial\Omega} Q(x,y)\mathrm{d}x + P(x,y)\mathrm{d}y.$$

This formula can be rewritten in a complex form. Let $F(z,\bar z) = F(x,y)$ be a (real) continuously differentiable function; then

$$\begin{aligned}\int_\Omega \partial_{\bar z} F(z,\bar z)\mathrm{d}x\mathrm{d}y &= \frac{1}{2}\int_\Omega [\partial_x F(x,y) + i\partial_y F(x,y)]\,\mathrm{d}x\mathrm{d}y \\ &= \frac{1}{2i}\int_{\partial\Omega} F(x,y)\mathrm{d}z. \qquad (1.1)\end{aligned}$$

In particular, if we use this formula with $F = \bar z$, we obtain the useful formula

$$\mathrm{Area}(\Omega) = \frac{1}{2i}\int_{\partial\Omega} \bar z \mathrm{d}z. \qquad (1.2)$$

Corollaries of the Cauchy formula. The Cauchy integral formula is the cornerstone of complex analysis and has numerous consequences. To a large extent, it is the reason why complex analysis is so different from real analysis. Here, we list some of its immediate corollaries that will be extensively used throughout the book.

Proposition 1.1. *Let f_n be analytic functions on a domain Ω. Let us assume that $f_n \to f$ locally uniformly on Ω. Then f is also an analytic function and $f'_n \to f$ locally uniformly on Ω.*

Theorem 1.2 (Liouville). *Let f be an entire function, that is, a function analytic in the entire complex plane $\mathbb{C}$. If f is bounded, then it must be a constant function.*

Proposition 1.3 (Maximum modulus principle). *Let f be analytic in Ω and $\bar{B}(z_0, r)$ be a closed disc inside Ω. Then*

$$|f(z_0)| \leq \sup_{\theta \in [0, 2\pi]} |f(z_0 + re^{i\theta})|.$$

The equality occurs if and only if f is constant in Ω. In particular, this implies that $|f|$ has no local maxima inside Ω unless f is constant.

Proofs of these three results are based on the Cauchy formula and can be found in many standard textbooks, in particular, in Rudin (1987, 10.23, 10.24, and 10.28).

Along the same lines, one can prove a similar result for harmonic functions.

Proposition 1.4 (Maximum principle for harmonic functions). *Let h be a harmonic function in Ω. Then h has no local extrema inside Ω. If h is continuous up to the boundary and $h \leq c$ on the boundary, then $h \leq c$ in Ω as well.*

Schwarz reflection. Schwarz reflection is the simplest way of extending certain analytic functions to analytic functions in larger domains. Although it can be applied only to a rather small class of functions, this method turned out to be very powerful and extremely useful.

Theorem 1.5 (Schwarz reflection principle). *Let Ω be a symmetric domain, that is, $z \in \Omega$ if and only if $\bar{z} \in \Omega$, Ω^+ be its upper half*

$$\Omega^+ = \{z \in \Omega : \operatorname{Im} z > 0\} = \Omega \cap \mathbb{H},$$

and L be the part of the real axis inside Ω. Suppose that f is a function analytic in Ω^+ and continuous in $\Omega^+ \cup L$. If $\operatorname{Im} f = 0$ on L, then f can be analytically extended to the entire Ω by $f(z) = \bar{f}(\bar{z})$.

The proof is based on the simple fact that $\bar{f}(\bar{z})$ is an analytic function in the lower half of Ω, and uses the Morera theorem to claim that the extension is analytic on L as well. The details can be found in many textbooks, in particular in Ahlfors (1978, Section 6.5). A similar result can be found in Rudin (1987, Theorem 11.14).

One way of approaching this result is in terms of symmetries, namely, transformations T such that $T \circ T$ is the identity. In our case, T is the symmetry with respect to the real line, namely $T(z) = \bar{z}$. Note that this function is anti-analytic (analytic as a function of $\bar{z}$). Schwarz reflection is based on the facts that if f is analytic, then $T \circ f \circ T$ is also analytic and that, since f is real on the real line, f and $T \circ f \circ T$ can be glued together to form a single analytic function. A similar argument can be used for other symmetries, in particular the symmetry with respect to a circle. The precise formulation is given by the following theorem.

Theorem 1.6 (Schwarz reflection principle). *Let Ω be a domain symmetric with respect to a circle $\{z : |z| = R\}$, that is, $z \in \Omega$ if and only if $T_R(z) = R^2/\bar{z} \in \Omega$. Let Ω^+ be its outer part, namely, $\Omega^+ = \{z \in \Omega, |z| > R\}$ and L be the part of the circle inside the domain, that is, $\Omega \cap \{z : |z| = R\}$. Suppose that f is a function analytic in Ω^+, continuous in $\Omega^+ \cup L$ and $|f| = r$ on L. Then f can be extended to the entirety of Ω by*

$$f(z) = T_r(f(T_R(z))) = r^2/\bar{f}(R^2/\bar{z}).$$

Argument principle. This is one of the first explicit examples of the connection between complex analysis and geometry. It shows that an analytic quantity (number of roots) is related to a geometric quantity (winding number). The standard statement is given by the following theorem.

Theorem 1.7 (Argument principle). *Let f be an analytic function in a domain Ω and γ be a positively oriented contractable simple closed curve in Ω. Then*

$$\frac{1}{2\pi i}\int_\gamma \frac{f'(z)}{f(z)}\mathrm{d}z$$

is equal to the number of zeroes of f inside γ (counting multiplicities). Alternatively, it is equal to the winding number (that is, the number of times $f(\gamma)$ goes around the origin) of $f(\gamma)$ with respect to the origin.

One of the standard corollaries of the argument principle is the Rouché theorem, which states that if two functions are close on a contour, then the number of zeroes inside is the same.

Theorem 1.8 (Rouché). *Let f and g be two analytic functions in some domain Ω and let $\gamma \subset \Omega$ be a closed contour. If $|f - g| < |f|$ on γ, then the functions f and g have the same number of zeroes inside γ.*

Proofs of the last two theorems can be found in Ahlfors (1978, Section 5.2).

Logarithm function. Very often, we will need to consider logarithms (all logarithms in this book will be natural) or powers of various functions. In many cases, the usual branch-cuts do not work, so we will need the following result.

Proposition 1.9. *Let Ω be a simply connected domain and f be an analytic function which does not vanish in Ω. Then it is possible to define a single-valued branch of* $\arg f$ *in Ω. In particular, this allows one to define single-valued branches of* $\log(f)$ *and* f^α.

This is a rather important result and there are two slightly different ways to approach it. We sketch the ideas behind both proofs.

The first proof is of geometric nature. It is well-known that for any simple curve that does not go through the origin, one can define a single-valued branch of argument along the curve. This is proved by covering the curve by a finite number of balls that do not contain the origin.

Let us fix a point $z_0 \in \Omega$ and a value of $\arg f(z_0) = \theta_0$. Let γ be a curve connecting z_0 and $z_1 \in \Omega$ inside Ω. By the previous argument, we can define a branch of argument on $f(\gamma)$ such that $\arg f(z_0) = \theta_0$. We define $\arg f(z_1)$ to be the value of the argument that is defined along $f(\gamma)$.

We have to show that this notion is well defined, namely, the value of the argument does not depend on a particular choice of γ.

Let us assume that this is not the case and there is another curve $\tilde{\gamma}$ connecting z_0 and z_1 such that the argument of $f(z_1)$ along $f(\tilde{\gamma})$ is different. By concatenating $f(\gamma)$ with the reverse of $f(\tilde{\gamma})$, we get a closed curve such that the increment of the argument along this curve is not zero. This implies that this curve has a non-trivial winding number, that is, it goes around the origin. Since the origin is not in $f(\Omega)$, this implies that this curve is not contractable inside $f(\Omega)$. On the other hand, this is an image of a closed curve inside a simply connected domain Ω. Since all such loops are contractable, its image must also be contractable within $f(\Omega)$, which contradicts our assumption (Figure 1.1).

This proves that our construction of $\arg f(z)$ is well defined. Since $\log f(z)$ and z^α can be defined in terms of the argument, this leads to the construction of these functions as well.

The second approach is more analytic. Let us fix some point $z_0 \in \Omega$ and consider the function

$$g(z) = \log(f(z_0)) + \int_{z_0}^{z} \frac{f'(z)}{f(z)} \mathrm{d}z,$$

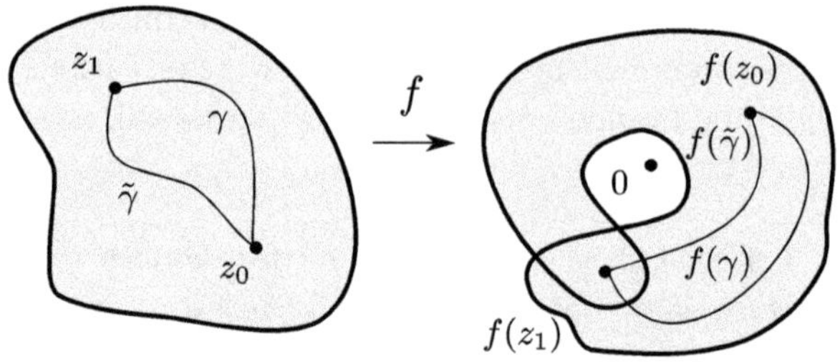

Fig. 1.1. The image of Ω might be non-simply connected, but the image of any (contractable) loop must be contractable.

where $\log(f(z_0))$ is any value of logarithm and the integral is along any curve connecting z_0 to z inside Ω. Since f'/f is analytic in a simply connected domain Ω, this integral is independent of the choice of the curve. The function g is a well-defined analytic function in Ω. It is not difficult to show that g is a branch of $\log(f)$.

Complex sphere. In many cases, it is convenient to work with functions that are analytic in the entire complex sphere $\widehat{\mathbb{C}}$ or in a domain which contains infinity. Although functions are not very different from the functions analytic in $\mathbb{C}$ or its subdomains, tweaking of several definitions is required. Functions analytic at infinity are often ignored in the basic complex analysis courses and books. Here, we give a very brief introduction.

We consider $\widehat{\mathbb{C}}$ to be the one-point compactification of $\mathbb{C}$, namely $\widehat{\mathbb{C}} = \mathbb{C} \cup \{\infty\}$. We will often think of infinity as a complex number. There is an obvious disadvantage of this approach: not all operations are defined. In particular, one cannot multiply infinity by zero or subtract infinity from itself.

It is possible to identify $\widehat{\mathbb{C}}$ with a sphere which is called the *complex sphere* or the *Riemann sphere*. One usually identifies infinity with the north pole and the remaining part of the sphere is identified with the complex plane $\mathbb{C}$ using the *stereographic projection*. Complex plane $\mathbb{C}$ is the horizontal plane passing through the equator, and each point z on the sphere (except the north pole) is projected to a point which is the intersection of the line passing through z and the north pole with the plane $\mathbb{C}$ (see Figure 1.2). More information about the Riemann sphere can be found in Gamelin (2001, Section 1.2). Under this identification, the unit disc $\mathbb{D}$ is the southern hemisphere, and $\mathbb{D}_- = \{z : |z| > 1\}$ is the northern hemisphere. It is important to note that here we presume $\mathbb{D}_-$ contains ∞.

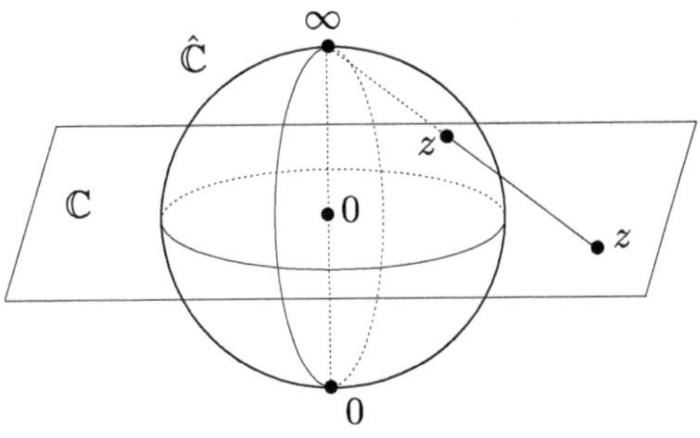

Fig. 1.2. Stereographic projection of the complex sphere onto the plane crossing the equator.

This projection allows one to define the *chordal distance* on the complex plane, namely, the distance in $\mathbb{C}$ which is the same as the Euclidean distance between the corresponding points on the unit sphere. It is not very difficult to compute the chordal distance between two points:

$$d(z,w) = \frac{2|z-w|}{\sqrt{1+|z|^2}\sqrt{1+|w|^2}},$$
$$d(z,\infty) = \frac{2}{\sqrt{1+|z|^2}}.$$

The infinitesimal form is

$$\mathrm{d}\sigma = \frac{2\mathrm{d}s}{1+|z|^2}.$$

This gives the *spherical metric* which corresponds to the lengths of curves on the sphere. In many cases, it is convenient to consider the (extended) complex plane equipped with this metric, since infinity is not an exceptional point with respect to this metric and many statements are easier to state and/or to prove.

This gives a natural topology in $\widehat{\mathbb{C}}$ and the corresponding notion of convergence. We say that $z_n \to \infty$ if for every $M > 0$, there is N such that $|z_n| > M$ for all $n > N$. With this definition, we can define the limits of functions at infinity and limits of functions that attain infinite values.

The definitions of analyticity are a bit more involved. The simplest is to use the mapping $z \mapsto 1/z$. Using this map, we can reformulate the continuity at infinity: $f(z)$ is continuous at infinity if $f(1/z)$ is continuous at zero; for a function such that $f(z_0) = \infty$, it is continuous at z_0 if $1/f(z)$ is continuous at z_0 (we naturally define $1/\infty$ to be 0). In the same spirit, we can deal with differentiability. A function f is differentiable at infinity if $f(1/z)$ is differentiable at 0. A function with $f(z_0) = \infty$ is differentiable at z_0 if $1/f(z)$ is differentiable.

We will see a lot of functions with $f(\infty) = \infty$. They are differentiable at infinity if $1/f(1/z)$ is differentiable at 0. Alternatively, for a function analytic in the neighbourhood of infinity, we can write a Laurent series

$$f(z) = \sum_{-\infty}^{\infty} a_n z^n, \qquad R < |z| < \infty.$$

The condition $f(\infty) = \infty$ is equivalent to the condition that not all a_n with $n > 0$ are equal to zero. If in addition we require f to be one-to-one near infinity, then it must be of the form

$$f(z) = a_1 z + a_0 + \sum_{n=1}^{\infty} a_{-n} z^{-n}.$$

For such functions, the derivative at infinity is defined as $f'(\infty) = a_1$.

Möbius transformations. Möbius transformations are often covered by the basic courses in complex analysis. Here, we provide a very short list of the main facts.

Möbius transformations are also called *linear fractional transformations.* They are functions of the form

$$f(z) = \frac{az+b}{cz+d},$$

where $a, b, c, d \in \mathbb{C}$ with $ad - bc \neq 0$. It is easy to see that the maps corresponding to the coefficients a, b, c, d and $\lambda a, \lambda b, \lambda c, \lambda d$ are the same provided that $\lambda \neq 0$.

These functions are analytic; moreover, they are bijective maps from $\widehat{\mathbb{C}}$ onto itself. This is one of the first examples motivating the use of $\widehat{\mathbb{C}}$. Later on, we will see that they are the only analytic maps of this type.

Any Möbius transformation maps circles and lines in $\mathbb{C}$ to circles and lines. This has a more natural form if one considers the Riemann sphere $\widehat{\mathbb{C}}$, since under the stereographic projections, circles and lines in $\mathbb{C}$ correspond to the circles in $\widehat{\mathbb{C}}$ (lines correspond to the circles passing through the north pole).

Direct computation shows that for any two triplets of distinct points (z_1, z_2, z_3) and (w_1, w_2, w_3), there is a unique Möbius transformation f such that $f(z_i) = w_i$. In particular, this implies that any circle in $\widehat{\mathbb{C}}$ can be mapped to any other circle.

For four points, this is no longer true. For a quadruplet, there is a non-trivial quantity which is preserved by the Möbius transformations: the *cross-ratio*

$$(z_1, z_2; z_3, z_4) = \frac{(z_1 - z_3)(z_2 - z_4)}{(z_2 - z_3)(z_1 - z_4)}.$$

This formula makes sense when all points $z_i \in \mathbb{C}$. If one of them is ∞, then the two factors involving it should be removed from the above formula. This quantity is invariant under the Möbius transformations, that is,

$$(z_1, z_2; z_3, z_4) = (f(z_1), f(z_2); f(z_3), f(z_4))$$

for any Möbius transformation f. Moreover, one quadruplet can be mapped onto another if and only if their cross-ratios are the same.

Sometimes, it is convenient to assume that the cross-ratio (z_1, z_2, z_3, z) is a function of z. Then it has a useful interpretation: as a function of z, this is the unique Möbius transformation which sends z_1 to ∞, z_2 to 0, and z_3 to 1.

Conformal maps. In this book, we are mostly interested in the one-to-one analytic functions. Since we think of them as mappings from one domain to another, we call them maps. It is a standard result of basic complex analysis that an analytic function f is locally one-to-one if and only if its derivative never vanishes. Such maps are called *conformal.* Slightly abusing notation, we will use this term for globally bijective analytic maps. It is easy to see that the condition that f' never vanishes does not imply global injectivity. Indeed, the function $f(z) = z^2$ is analytic in the complement of the unit disc and its derivative does not vanish there, but it is a two-to-one map. There are two other terms for analytic one-to-one maps: *univalent* and *schlicht.* We use these terms interchangeably.

Chapter 2

Riemann Mapping Theorem

In this chapter, we discuss a class of results that form a foundation for the rest of the book. We are interested in conformal classification of planar domains, that is, given two domains, can we find a conformal map from one domain onto the other?[1]

There is an obvious topological obstacle. Since conformal maps are bicontinuous bijections, conformally equivalent domains are also topologically equivalent. This means that there is no way to map, say, a simply connected domain onto a doubly connected domain.

One would expect that conformal equivalence is significantly more restrictive than a topological equivalence. This is indeed the case, but there are some surprises; in particular, almost all simply connected domains are conformally equivalent to each other. We start with the complete study of the simply connected case and then briefly discuss more complicated domains of higher connectivity.

In this chapter, we discuss a class of uniformizing results. We are mostly interested in the following broad question: given a domain in the complex plane, can we find a conformal map from this domain onto some simple domain? The first result in this direction is the famous Riemann mapping theorem, which states that any simply connected domain can be conformally mapped onto the complex sphere $\widehat{\mathbb{C}}$, the complex plane $\mathbb{C}$, or the unit disc $\mathbb{D}$.

[1]There is a very closely related Riemann uniformization theorem which deals with the same question for Riemannian surfaces, but it is not the subject of this book.

We will present the classical proof by Fejer and Riesz. We will briefly mention some other approaches to the construction of the uniformizing maps and the Riemann mapping theorem for multiply connected domains.

2.1. Historical Remarks

The Riemann mapping theorem first appeared in Riemann's inaugural dissertation in 1851 (Riemann, 1851). He stated:[2]

> Two given simply connected plane surfaces can always be related in such a way that each point of one surface corresponds to a point of the other, varying continuously with that point, with the corresponding smallest parts similar. One interior point, and one boundary point, can be assigned arbitrary corresponding points; however, this determines the correspondence for all points.

It is important to note that for Riemann, the domain was a planar set bounded by a curve, which does not cover all the domains that we will consider in this book. For us, the domain is any simply connected open set. His proof was based on the connection between the existence of the mapping between the domain and the disc and the existence of Green's function. We will discuss this connection in detail in Section 2.8. To show the existence of Green's function, Riemann used the Dirichlet principle: the function u solving the Dirichlet boundary problem in a domain Ω is the unique minimizer of the Dirichlet integral

$$D(u) = \iint_\Omega (\partial_x u)^2 + (\partial_y u)^2 \mathrm{d}x\mathrm{d}y = \iint_\Omega |\nabla u|^2.$$

It was very soon pointed out by Weierstrass that the existence of the non-trivial lower bound on $D(u)$ does not mean that there is a minimizer. It also became clear that the boundary of a domain could be more complicated than what had been assumed by Riemann.

In 1887, Harnack published a book where he used the approximation of a general simply connected domain by polygonal domains to solve the Dirichlet boundary problem. Osgood pointed out that the boundary of a simply connected domain could be even more complicated than Harnack assumed and his method could fail for domains with such complicated boundaries. Building upon the ideas of Harnack and Poincaré, he gave in Osgood (1900)

[2]Quoted from an English translation (Riemann, 2004).

the first complete proof of the existence of Green's function for general planar simply connected domains different from the entire plane. He also pointed out that if the boundary is not a simple Jordan curve, then the boundary values might be undefined and the normalization used by Riemann might not make sense.

Later on, based on the previous work of Carathéodory, Koebe (1912) gave a proof based on function theory methods rather than on potential theory. His method is very similar to the one presented in Section 2.6, and is closely related to the proof presented in Section 2.4, which was first obtained by Fejer and Riesz and presented by Radó (1923). More on the history of the proof can be found in Gray (1994) and Walsh (1973).

2.2. Möbius Transformations and the Schwarz Lemma

Before proving the existence of the map from a given domain onto a standard domain, we would like to discuss its uniqueness assuming its existence.

Let Ω be a domain in the complex sphere and let us assume that there are two conformal maps f and g from Ω onto some uniformizing domain Ω'. Then the map $\mu = g \circ f^{-1}$ is a conformal automorphism of Ω', that is, a conformal transformation from Ω' onto itself. Conversely, if μ is an automorphism of Ω', then $\mu \circ f$ is also a conformal map from Ω onto Ω'. This means that the non-uniqueness of f is given by the collection of all conformal maps of Ω' onto itself. It is important to note that this collection forms a group with respect to the composition. It is called the group of conformal automorphisms of Ω'.

In this book, we often work with the 'standard' domains $\widehat{\mathbb{C}}$, $\mathbb{C}$, $\mathbb{D}$, and $\mathbb{H} = \{z : \operatorname{Im} z > 0\}$. Later on, we see that the first three cannot be mapped conformally onto each other, but it is easy to construct a conformal map between the last two. In fact, this map is a Möbius transformation.

Exercise 2.1. Find Möbius transformations between $\mathbb{D}$ and $\mathbb{H}$.

In this section, we describe all conformal automorphisms of $\widehat{\mathbb{C}}$, $\mathbb{C}$, $\mathbb{H}$, and $\mathbb{D}$. It is well known that there are Möbius transformations preserving these domains. Any Möbius transformation is a conformal automorphism of $\widehat{\mathbb{C}}$. For the other domains, the Möbius transformations preserving them are described by the following proposition.

Proposition 2.1. *The only Möbius transformations that map $\mathbb{D}$, $\mathbb{C}$, or $\mathbb{H}$ to themselves are of the form*

$$\begin{aligned} &f:\mathbb{D}\to\mathbb{D}, && f(z)=e^{i\theta}\frac{z-a}{1-\bar{a}z}, && a\in\mathbb{D},\ \theta\in\mathbb{R},\\ &f:\mathbb{C}\to\mathbb{C}, && f(z)=az+b, && a,b\in\mathbb{C},\\ &f:\mathbb{H}\to\mathbb{H}, && f(z)=\frac{az+b}{cz+d}, && a,b,c,d\in\mathbb{R},\ ad-bc>0. \end{aligned}$$

Exercise 2.2. Prove Proposition 2.1.

It turns out that these Möbius transformations are the *only* conformal automorphisms of these domains. To prove this, we will need a classical result, known as the Schwarz lemma. It is a rather elementary but very powerful result, which we will use extensively.

Theorem 2.2 (Schwarz lemma). *Let f be an analytic function in the unit disc $\mathbb{D}$ normalized to have $f(0)=0$ and $|f(z)|\le 1$. Then $|f(z)|\le|z|$ for all $z\in\mathbb{D}$ and $|f'(0)|\le 1$. Moreover, if $|f(z)|=|z|$ for some $z\ne 0$ or $|f'(0)|=1$, then $f(z)=e^{i\theta}z$ for some $\theta\in\mathbb{R}$.*

Proof. Let us define $g(z)=f(z)/z$ for $z\neq 0$. Since f has zero at the origin, g has a removable singularity: it is analytic in $\mathbb{D}$ if we define $g(0)=f'(0)$. Next, let us fix some $0<r<1$. On the circle $|z|=r$, we have $|g(z)|<1/r$ and hence, by the maximum modulus principle, the same is true for $|z|<r$. Passing to the limit as $r\to 1$, we show that $|g|\le 1$ in $\mathbb{D}$, which is equivalent to $|f(z)|\le|z|$ and $|f'(0)|\le 1$.

Now, assume that there is a point inside $\mathbb{D}$ where $|g(z)|=1$. By the maximum modulus principle, g must be a constant of modulus one, equivalently $g(z)=e^{i\theta}$ for some real θ. This proves the second part of the theorem. □

Note that the normalization that we use is not very restrictive: by rescaling and adding a constant, any bounded function in $\mathbb{D}$ can be reduced to this form.

One way of approaching the Schwarz lemma is to compare f with a Möbius transformation $\mu:\mathbb{D}\to\mathbb{D}$ such that $\mu(0)=0$. It turns out that this can be generalized to other Möbius transformations.

Theorem 2.3. *Let f be an analytic function in $\mathbb{D}$ such that $|f|\le 1$. Then for any $z,w\in\mathbb{D}$ we have*

$$\left|\frac{f(w)-f(z)}{1-\bar{f}(z)f(w)}\right|\le\left|\frac{w-z}{1-\bar{z}w}\right|$$

and

$$|f'(z)| \leq \frac{1-|f(z)|^2}{1-|z|^2}.$$

Moreover, equality happens only if f is a Möbius transformation preserving the unit disc.

Proof. Let us fix z and consider a Möbius transformation $h(\zeta) = (\zeta + z)/(1 + \bar{z}\zeta)$. The function

$$F(\zeta) = \frac{f(h(\zeta)) - f(h(0))}{1 - \bar{f}(h(0))f(h(\zeta))}$$

is a composition of f with two Möbius transformations that preserve $\mathbb{D}$. It is also obvious from the definition that $F(0) = 0$ and thus it satisfies all assumptions of the Schwarz lemma; this implies that $|F(z)| \leq |z|$ and $|F'(0)| \leq 1$. To complete the proof, we note that $\zeta = (w - z)/(1 - \bar{z}w)$ is equal to $h^{-1}(w)$. Plugging this value of ζ into the inequalities for $|F(\zeta)|$ and $|F'(0)|$, we complete the proof of the theorem. □

Finally, we prove one more inequality which also follows immediately from the Schwarz lemma.

Lemma 2.4. *Let $\Omega_1 \subset \Omega_2$ be two simply connected domains and let $f_k : \Omega_k \to \mathbb{D}$ be two conformal maps such that $f_k(z_0) = 0$ for some fixed $z_0 \in \Omega_1$. We assume that there is a conformal map $\phi : f_2(\Omega_1) \to \mathbb{D}$ such that $f_1(z) = \phi(f_2(z))$. Then, for all $z \in \Omega_1$,*

$$|f_1(z)| \geq |f_2(z)|.$$

In this lemma, we assumed the existence of functions f_1, f_2, and ϕ; later on we will see that, by the Riemann mapping theorem, such functions always exist.

Proof of Lemma 2.4. The proof is a direct application of the Schwarz lemma. Indeed, function ϕ^{-1} satisfies the assumptions of the Schwarz lemma; hence $|\phi^{-1}(z)| \leq |z|$, or, equivalently, $|z| \leq |\phi(z)|$. Combining this with $f_1(z) = \phi(f_2(z))$, we prove the lemma. □

Finally, we can use the Schwarz lemma to prove that all conformal automorphisms of standard domains are Möbius transformations.

Proposition 2.5. *All conformal automorphisms of $\widehat{\mathbb{C}}$, $\mathbb{C}$, $\mathbb{H}$, and $\mathbb{D}$ are Möbius transformations.*

Proof. We prove the unit disc case; the other cases are left as exercises.

Let $f : \mathbb{D} \to \mathbb{D}$ be a conformal automorphism. We define a Möbius transformation $\mu(z) = (z - w)/(1 - \bar{w}z)$, where $w = f(0)$. Obviously, $g = \mu(f)$ is a conformal automorphism of $\mathbb{D}$ with $g(0) = 0$ and $|g(z)| \leq 1$. By the Schwarz lemma, $|g(z)| \leq |z|$. But we can also apply the Schwarz lemma to the inverse map g^{-1} and obtain $|g^{-1}(z)| \leq |z|$. This means that $|g(z)| = |z|$ and hence $g(z) = e^{i\theta}z$ for some θ. This proves that f is an inverse of the Möbius transformation $e^{-i\theta}\mu(z)$, and hence it is also a Möbius transformation. □

Exercise 2.3. Prove the Proposition 2.5 for $\widehat{\mathbb{C}}$, $\mathbb{C}$, and $\mathbb{H}$.

2.3. Normal Families

In this section, we discuss some results of the convergence of conformal maps that we will need for the proof of the Riemann mapping theorem.

Definition 2.6. Let $\mathcal{F}$ be a family of analytic functions on Ω. We say that $\mathcal{F}$ is a *normal family* if, for every sequence f_n of functions from $\mathcal{F}$, there is a subsequence which converges uniformly on all compact subsets of Ω. This type of convergence is called *normal convergence.*

The term 'normal family' is somewhat old-fashioned; in more modern terms, it should be called 'precompact'.

Exercise 2.4. Consider a family of analytic maps $f_n : \widehat{\mathbb{C}} \to \widehat{\mathbb{C}}$ given by $f_n(z) = z^n$. Show that this family is normal in $\mathbb{D}$ but not in any domain containing a point of modulus 1.

Exercise 2.5. Let $\mathcal{F}$ be the family of all functions of the form $f_w(z) = z/(z - w)$ with $|w| > 1$. Give a direct proof that this family is normal in the unit disc.

The standard way to prove precompactness is to use the Arzelà–Ascoli theorem, and this is exactly what we will do. Before stating the theorems, we need two more definitions.

Definition 2.7. We say that a family of functions $\mathcal{F}$ defined on Ω is *equicontinuous* in $A \subset \Omega$ if for every $\varepsilon > 0$, there is $\delta > 0$ such that $|f(x) - f(y)| < \delta$ for every $f \in \mathcal{F}$ and all $x, y \in A$ such that $|x - y| < \varepsilon$.

Definition 2.8. We say that a family $\mathcal{F}$ of functions defined on Ω is *uniformly bounded* on $A \subset \Omega$ if there exists M such that $|f(x)| < M$ for all $x \in A$ and every $f \in \mathcal{F}$.

Now we can state the Arzelà–Ascoli theorem, which we present here without proof. Interested readers can find it in many books, including Rudin (1987, Theorem 11.28) and Gamelin (2001, Section XI.5).

Theorem 2.9 (Arzelà–Ascoli). *Let $\mathcal{F}$ be a family of pointwise bounded equicontinuous functions from a separable metric space*[3] *X to $\mathbb{C}$. Then every sequence f_n of functions from $\mathcal{F}$ contains a subsequence that converges uniformly on all compact subsets of X.*

Now we are ready to state and prove Montel's theorem, which gives a simple sufficient condition for normality of a family of analytic functions.

Theorem 2.10 (Montel). *Let $\mathcal{F}$ be a family of analytic functions on a domain Ω that is uniformly bounded on every compact subset of Ω. Then $\mathcal{F}$ is a normal family.*

Proof. First, we construct a family of compacts that exhaust Ω. We define K_n to be the set of all $z \in \Omega$ such that $|z| \le n$ and $\operatorname{dist}(z, \mathbb{C} \setminus \Omega) \ge 1/n$. (We assume that all $K_n \neq \emptyset$, otherwise we change indexes so that K_1 is the first non-empty set.) It is easy to see that for every compact $K \subset \Omega$, there is n such that $K \subset K_n$. This also implies that $\bigcup K_n = \Omega$. Moreover, K_n are increasing and separated, namely $K_n \subset K_{n+1}$ and there are $\delta_n > 0$ such that $B(z, \delta_n) \subset K_{n+1}$ for all $z \in K_n$.

Let z and w be two points from K_n with $|z - w| < \delta_n/2$ and f be any function from $\mathcal{F}$. We can use the Cauchy formula to write

$$f(z) - f(w) = \frac{1}{2\pi i} \int_\gamma \left(\frac{1}{\zeta - z} - \frac{1}{\zeta - w} \right) f(\zeta) \mathrm{d}\zeta,$$

where γ is a circle of radius δ_n centred at z. Note that $\gamma \subset K_{n+1}$ and, since $\mathcal{F}$ is uniformly bounded, there is a constant M_{n+1} independent of f such that $|f(\zeta)| \le M_{n+1}$. This allows us to estimate

$$|f(z) - f(w)| \le \frac{2M_{n+1}}{\delta_n} |z - w|,$$

[3]There is no need to know anything about separable metric spaces, not even the definitions. We give the statement in a rather general form, but we will use it only in the case where X is a subset of $\mathbb{C}$, in which case it is separable and metric. For our purposes, it is sufficient to know that the Arzelà–Ascoli theorem is applicable for subsets of $\mathbb{C}$.

which implies that $\mathcal{F}$ is equicontinuous on K_n and hence on every compact subset of Ω.

By the Arzelà–Ascoli theorem, from each sequence of functions f_n from $\mathcal{F}$ we can choose a subsequence converging uniformly on K_n. Let $f_{1,n}$ be a subsequence converging on K_1. By the same argument, it has a subsequence converging on K_2, which we denote by $f_{2,n}$. Continuing in this way, we construct a family of sequences $f_{k,n}$. By the standard diagonal argument, the sequence $f_{n,n}$ converges uniformly on every K_n and hence on every compact subset of Ω. □

Exercise 2.6. Let $\mathcal{F}$ be the family of all analytic functions in $\mathbb{D}$ such that $\operatorname{Re} f > 0$ and $f(0)$ is uniformly bounded. Show that $\mathcal{F}$ is normal.

Exercise 2.7. We define the Hardy space $H^1 = H^1(\mathbb{D})$ as the space of all holomorphic functions in $\mathbb{D}$ such that

$$\|f\|_{H^1} = \sup_{0<r<1} \frac{1}{2\pi} \int_0^{2\pi} |f(re^{i\theta})| d\theta < \infty.$$

Show that the family of functions $f \in H^1$ such that $\|f\|_{H^1} \leq 1$ is normal.

One of the useful corollaries of the Montel theorem is the following result, known as the Vitali or Vitali–Porter theorem. Here, we derive it from Montel's theorem. Alternatively, it can be proved directly and Montel's theorem derived from it.

Theorem 2.11 (Vitali). *Let f_n be a sequence of analytic functions on a domain Ω that is uniformly bounded on each compact subset of Ω. Then the functions f_n converge to f uniformly on compact subsets of Ω if and only if there is a set of points A, such that A has a point of accumulation in Ω and f_n converge pointwise on A.*

Proof. One direction is trivial: if $f_n \to f$ uniformly on compacts, then the set A is equal to Ω and clearly has accumulation points.

Let us assume that A has a point of accumulation, that is, there are distinct $a_n \in A$ such that $a_n \to a \in \Omega$. By Montel's theorem, the family $\{f_n\}$ is normal, which means that there are subsequential limits. If all subsequential limits are equal, then there is a limit and there is nothing to prove. Let us assume that there are two subsequences $f_{1,n}$ and $f_{2,n}$ which converge to two different limits F_1 and F_2. Since convergence is uniform on compacts, both F_1 and F_2 are also analytic in Ω.

Let a be any point of A. Since both sequences are subsequences of f_n and $f_n(a)$ converges, we have

$$F_1(a) = \lim_{n\to\infty} f_{1,n}(a) = \lim_{n\to\infty} f_n(a) = \lim_{n\to\infty} f_{2,n}(a) = F_2(a).$$

This means that two analytic functions are equal on the set A which has a point of accumulation. By the standard identity theorem, this implies that functions are equal everywhere. This is one of the manifestations of 'rigidity' of analytic functions. □

This result is a classical example of the *propagation of convergence*: convergence on a small set implies convergence everywhere. This is one of the manifestations of 'rigidity' of analytic functions.

It is important to mention that Montel's theorem tells us very little about the limit of the subsequence. From uniform convergence, we know that the limit is also analytic in Ω, but we do not know whether it belongs to $\mathcal{F}$ or not. We are mostly interested in the case when all functions from $\mathcal{F}$ are univalent; in this case, we have the following dichotomy.

Theorem 2.12 (Hurwitz). *Let f_n be a sequence of univalent functions in Ω that converge uniformly on compacts to f. Then f is either a univalent or a constant function.*

Remark 2.13. This is a typical example of a dichotomy in complex analysis where we can say that our object is either as good as possible or as bad as possible, but not something in between. Hurwitz's theorem states that the limit of univalent maps is either univalent or as non-univalent as a function could be, that is, it is a constant.

Another example is the classification of isolated singularities: either a function has a limit in $\widehat{\mathbb{C}}$, in which case it is a removable singularity or a pole, or every value is a subsequential limit, in which case it is an essential singularity.[4] Again, we have a result that either there is a limit, or the function is as far from being continuous as it could be.

Proof of Theorem 2.12. Let us assume that the limiting function f is not univalent, that is, there are distinct points z_1 and z_2 in Ω such that $f(z_1) = f(z_2)$. The sequence of functions $g_n(z) = f_n(z) - f_n(z_2)$

[4]In fact, an even stronger result holds. Picard's great theorem states that if an analytic function f has an essential isolated singularity at z_0, then in any neighbourhood of z_0, the function f assumes every value with one possible exception.

converges to $g(z) = f(z) - f(z_2)$. Let us assume that f is not a constant function; then the roots of g are isolated and there is a small circle γ around z_1 such that $\gamma \subset \Omega$, g does not vanish on γ and z_2 is not inside γ. Since g does not vanish on γ, there is $c > 0$ such that $|g| > c$ on γ. By uniform convergence, $|g - g_n| < c$ on γ for sufficiently large n. By Rouché's theorem (Theorem 1.8), the numbers of the roots of g and g_n inside of γ are the same for sufficiently large n. But functions g_n are univalent and $g(z_2) = 0$, and hence there are no roots inside γ, but $g(z_1) = 0$. This contradiction implies that if f is not univalent, so it must be constant. □

2.4. Proof of the Riemann Mapping Theorem

We start with the complete statement of the Riemann mapping theorem, which is rather different from the original statement.

Theorem 2.14. *Let Ω be a simply connected domain in the complex sphere $\widehat{\mathbb{C}}$. Then Ω is conformally equivalent to one of the three domains: $\widehat{\mathbb{C}}$, $\mathbb{C}$, or $\mathbb{D}$. To be more precise, if $\widehat{\mathbb{C}} \setminus \Omega$ contains at least two points, then Ω is equivalent to $\mathbb{D}$; if it contains one point, then it is equivalent to $\mathbb{C}$; and if it is empty, then $\Omega = \widehat{\mathbb{C}}$.*

Moreover, if Ω is equivalent to $\mathbb{D}$ and z_0 is any point in Ω, then there is a unique conformal map $f : \Omega \to \mathbb{D}$ such that $f(z_0) = 0$ and $f'(z_0) > 0$. (Here and later on, when we write that some complex quantity is positive, we mean that it is real and positive. This is also equivalent to the statement that the argument is 0.)

The three uniformizing domains $\widehat{\mathbb{C}}$, $\mathbb{C}$, and $\mathbb{D}$ are not conformally equivalent.

Proof. We start from the last part of the theorem. It is easy to see that $\widehat{\mathbb{C}}$ cannot be equivalent to $\mathbb{C}$ or $\mathbb{D}$ since they are not even homeomorphic. To show that $\mathbb{C}$ and $\mathbb{D}$ are not equivalent, we assume the contrary, that is, that there is a univalent map from $\mathbb{C}$ onto $\mathbb{D}$. This function is a bounded entire function. By Liouville's theorem, this function must be constant, which contradicts our assumption that it is univalent.

There is nothing to prove when $\Omega = \widehat{\mathbb{C}}$. When $\Omega = \widehat{\mathbb{C}} \setminus \{w_0\}$, we can apply the Möbius transformation $\mu = 1/(z - w_0)$, which maps Ω onto $\mathbb{C}$.

The only interesting case is when the complement of Ω contains at least two points. To analyze this case, we consider the family $\mathcal{F}$ of all univalent maps f on Ω such that $|f(z)| \leq 1$, $f(z_0) = 0$, and $f'(z_0) > 0$ for some fixed $z_0 \in \Omega$.

We will take the following steps to complete the proof:

(1) Show that the family $\mathcal{F}$ is non-empty.
(2) Show that the family $\mathcal{F}$ is normal.
(3) Consider a continuous functional on $\mathcal{F} : f \mapsto f'(0)$. Show that there is $f \in \mathcal{F}$ which maximizes $f'(0)$.
(4) Show that the maximizer is a conformal map onto $\mathbb{D}$.

Step 1. We know that there are two points outside of Ω. By applying a Möbius transformation, we can assume that one of these points is infinity and the other is the origin. So, our domain is a proper simply connected subdomain of $\mathbb{C}$. Since Ω is simply connected, there is a branch cut connecting 0 to infinity that lies outside of Ω. Using this branch cut, we can define a single-valued branch of $\phi(z) = z^{1/2}$ in Ω. Note that this function is univalent. Indeed, if $\phi(z_1) = \phi(z_2)$, then $z_1 = z_2$. By the same argument, it does not take the opposite values, that is, we cannot have $\phi(z_1) = -\phi(z_2)$. Since ϕ maps a small neighbourhood of z_0 onto an open neighbourhood of $w_0 = \phi(z_0)$, there is $r > 0$ such that $B(w_0, r) \subset \phi(\Omega)$ and $B(-w_0, r) \cap \phi(\Omega) = \emptyset$. Composing ϕ with $r/(z + w_0)$, we find a map from $\mathcal{F}$.

Note that for domains with non-empty interior of the complement, we only need the last step. The trick with the square root is needed only for domains that are dense in $\mathbb{C}$.

Step 2. Since all functions in $\mathcal{F}$ are bounded by 1, normality follows immediately from Montel's theorem (Theorem 2.10).

Step 3. It is a standard corollary of the Cauchy formula that if analytic functions f_n converge uniformly to f, then $f_n'(z) \to f'(z)$ for every z. This proves that the functional $f \mapsto f'(z_0)$ is continuous with respect to the uniform convergence on compact sets.

Let M be the supremum of $f'(z_0)$ over all functions from $\mathcal{F}$. There is a sequence f_n such that $f_n'(z_0) \to M$ (note that we do not assume that M is finite). By the normality of $\mathcal{F}$, there is a subsequence which converges on all compact subsets of Ω. Abusing notation, we denote this subsequence by f_n and its limit by f. Uniform convergence implies that f is analytic in Ω and $f'(z_0) = M$. In particular, M is finite.

By Hurwitz's theorem (Theorem 2.12), the limit f is either univalent or constant. Since $M > 0$, f cannot be constant.

Step 4. The main idea of this step is rather simple. In some sense, the derivative at z_0 controls how far the images of other points are from $f(z_0)$.

If there is a point w in $\mathbb{D} \setminus f(\Omega)$, then we can construct a function that will push w towards the boundary of $\mathbb{D}$. Explicit computation will show that the composition of f with this function has a larger derivative.

First, we compose f with a Möbius transformation $\mu(z) = (z - w)/(1 - \bar{w}z)$. This will map w to the origin. Now, by the same argument as in the first step, we can define a single-valued branch of

$$F(z) = (\mu(f(z)))^{1/2} = \sqrt{\frac{f(z) - w}{1 - \bar{w}f(z)}}.$$

Finally, we have to compose F with another Möbius transformation that will send $F(z_0)$ back to the origin. This is done by

$$G(z) = \frac{|F'(z_0)|}{F'(z_0)} \frac{F(z) - F(z_0)}{1 - F(z)\overline{F(z_0)}}.$$

The first factor is needed to ensure that the derivative at z_0 is positive (in other words, its argument is zero). An explicit computation shows that

$$G'(z_0) = \frac{|F'(z_0)|}{1 - |F(z_0)|^2} = \frac{1 + |w|}{2\sqrt{|w|}} f'(z_0) > f'(z_0).$$

By construction, $G \in \mathcal{F}$ and $G'(z_0) > f'(z_0)$; this contradicts the assumption that f maximizes the derivative at z_0.

To complete the proof of the theorem, we have to show that the map f is unique. Let us assume that there is another function g which maps Ω onto $\mathbb{D}$ and has the right normalization. The map $f \circ g^{-1}$ is a conformal automorphism of the unit disc. By Proposition 2.1, it has the form

$$e^{i\theta} \frac{z - a}{1 - \bar{a}z}.$$

Since 0 is mapped to itself and the derivative at 0 is positive, we must have $a = 0$ and $e^{i\theta} = 1$. This means that $g^{-1} = f^{-1}$ and $f = g$. □

We can see from the proof that the univalent map onto the disc maximizes the derivative at the point which is mapped to the origin. There is an alternative extremal formulation. Let us assume that Ω is a simply connected domain such that $\widehat{\mathbb{C}} \setminus \Omega$ contains at least two points. We denote by $\mathcal{F}$ the family of all univalent maps on Ω with $f(z_0) = 0$ and $f'(z_0) = 1$. The functional $f \mapsto \sup |f(z)|$ is minimized by the unique univalent map onto the disc of radius

$$R = \min_{f \in \mathcal{F}} \sup_{z \in \mathbb{D}} |f(z)|.$$

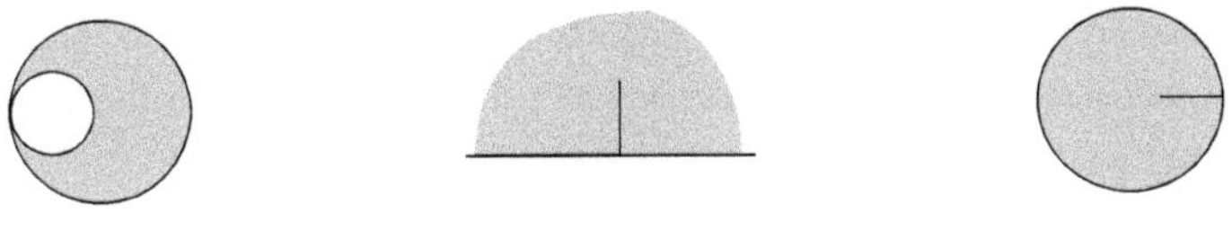

(a) (b) (c)

Fig. 2.1. Three examples of domains for which the uniformizing maps can be found explicitly: (a) a digon between two tangent circles; (b) the upper half-plane with a vertical slit; (c) the unit disc with a radial slit.

This radius is called the *conformal radius of the domain* Ω *at* z_0 and is denoted by $\rho(z_0, \Omega)$. Alternatively, the conformal radius could be defined as $|F'(0)|$ where F is a univalent map from $\mathbb{D}$ onto Ω such that $F(0) = z_0$.

There is one more statement claiming that the derivative at a fixed point is related to the size of the domain. This result is known as Lindelöf's principle. Let f_1 and f_2 be two univalent functions mapping $\mathbb{D}$ onto Ω_1 and Ω_2, respectively. We also assume that $f_i(0) = 0$ and that $\Omega_1 \subset \Omega_2$. Then $|f_1'(0)| \le |f_2'(0)|$ with equality holding if and only if $f_2(z) = f_1(e^{i\theta}z)$ for some real θ.

Minimization of the maximum modulus and Lindelöf's principle follow immediately from the proof of the Riemann mapping theorem. Lindelöf's principle also implies that the conformal radius increases when the domain increases.

Exercise 2.8. Find a conformal map onto $\mathbb{D}$ or $\mathbb{H}$ for the following domains:

(1) An infinite strip $S = \{z : 0 < \operatorname{Im} z < 1\}$.
(2) A domain bounded by two touching circles, namely, let Ω be the domain between two circles of radii r and R centred at r and R (see Figure 2.1(a)).
(3) The upper half-plane with a slit $\Omega = \mathbb{H} \setminus [0, it]$ with $t > 0$ (see Figure 2.1(b)).
(4) The unit disc with a slit $\Omega = \mathbb{D} \setminus [x, 1]$ with $-1 < x < 1$ (see Figure 2.1(c)).

2.5. Other Normalizations

The Riemann mapping theorem (Theorem 2.14) shows that all simply connected domains whose complement contains at least two points are conformally equivalent. In the proof of this theorem, we used the unit disc as the standard uniformizing domain. Obviously, this choice is completely arbitrary. In this section, we discuss other uniformizing domains and normalizations.

First, we know that the map from a simply connected domain Ω onto $\mathbb{D}$ is not unique; it can be composed with any Möbius transformation preserving the unit disc. The family of these transformations is described by three real parameters: the real and imaginary parts of the point which is mapped to the origin and the angle of rotation. This means that, in general, we should be able to fix uniquely any three real parameters by the proper choice of a Möbius transformation.

In the standard formulation of the Riemann theorem, we normalize the map by requiring that a fixed point z_0 is mapped to the origin and that the argument of the derivative at this point is zero. This corresponds exactly to fixing three real parameters, so it should not be a surprise that such a map is unique. We would like to point out that the argument with the number of parameters is just a rule of thumb, although a very good one, and each separate case requires a rigorous proof.

Other standard ways to choose normalization are as follows: fix one interior and one boundary point (as in the original Riemann's statement of the Riemann mapping theorem), fix three boundary points, and fix two boundary points and the derivative at one of them. For some of these normalizations, other domains are natural uniformizing domains.

All these normalizations use f or even f' on the boundary which, as we will see later, might not exist. However, they do exist for domains with sufficiently 'nice' boundaries. We postpone the discussion of the existence of the boundary values until Section 2.7. Here, we just assume that all boundary values and derivatives that we use are well defined.

One interior and one boundary point. This is the normalization that appeared in the original statement of the Riemann theorem. Let Ω be a domain conformally equivalent to $\mathbb{D}$ and let f be a conformal map from Ω onto $\mathbb{D}$. We choose an interior point $z_0 \in \Omega$ and a boundary point $\zeta \in \partial\Omega$. We assume that f can be defined continuously at ζ. Then there is a unique univalent function $g : \Omega \to \mathbb{D}$ such that $g(z_0) = 0$ and $g(\zeta) = 1$.

By the Riemann theorem, we can assume that $f(z_0) = 0$ and we know that all maps onto $\mathbb{D}$ differ by composition with a Möbius transformation. By the Schwarz lemma, the only Möbius automorphisms of $\mathbb{D}$ preserving the origin are rotations. This means that $f(z)/f(\zeta)$ is the only map with the desired properties.

Three boundary points. As before, we assume that there is a map $f : \Omega \to \mathbb{D}$ which can be continuously defined at boundary points ζ_i, $i = 1, 2, 3$. Let z_i be three points on the boundary of $\mathbb{D}$ that have the same order as ζ_i.[5] We know that there is a unique Möbius transformation μ mapping $f(\zeta_i)$ to z_i. Note that μ will also map the unit circle to itself. Since z_i and $f(\zeta_i)$ have the same order, map μ will send the unit disc to itself (if they are in the opposite order, then $\mathbb{D}$ is mapped onto $\mathbb{D}_-$). This means that $\mu \circ f$ will send ζ_i to z_i.

Sometimes, the unit disc is not the most convenient domain for this type of normalization. It can be more useful to map Ω onto the upper half-plane and to send the three given points to 0, 1, and ∞.

Two boundary points and derivative. First, we have to assume that the function $f : \Omega \to \mathbb{D}$ is continuous at the two boundary points ζ_1 and ζ_2. Here we have to assume not only that f could be extended continuously to the boundary, but also that $f'(\zeta_1)$ makes sense. We do not want to discuss this condition in detail, but would like to mention that this is true if the boundary of Ω near ζ_1 is an analytic curve.

It might seem that we want to fix too many parameters: the two boundary points give us two real parameters and a derivative is a complex number; hence it also gives two parameters. But we can note that near ζ_1 the function f maps the smooth boundary of Ω onto smooth boundary of the unit disc. This determines the argument of the derivative and we are left with only one free parameter: the modulus of the derivative.

The best uniformizing domain for this problem is the half-plane. As before, by composing with a Möbius transformation, we can construct a map $g : \Omega \to \mathbb{H}$ with $g(\zeta_1) = 0$ and $g(\zeta_2) = \infty$. It is easy to see that $g(z)/|g'(\zeta_1)|$ maps ζ_1 and ζ_2 to 0 and ∞ and has a derivative of modulus 1 at ζ_1. It is easy to check that we can choose any two points and the value of the modulus of the derivative, but this particular normalization is probably the most useful one.

Exercise 2.9. Prove that this map f is unique.

Thermodynamic normalization. Let us consider a case when $\Omega = \mathbb{H} \setminus K$ where K is a compact subset of $\mathbb{H} \cup \mathbb{R}$ and Ω is a simply connected domain.

---

[5] For domains with nice boundaries, it is clear what it means for points to be in, say, clockwise order. For more complicated domains, we could define the order on the boundary of Ω by the order of the images under the Riemann map.

We usually assume that the closure of $K \cap \mathbb{H}$ is equal to K, which basically means that K is the 'minimal' set with given $\mathbb{H} \setminus K$. But sometimes, it is convenient to add a real interval to K. This changes K, but not the corresponding domain $\mathbb{H} \setminus K$. Such sets K are called (*compact*) *half-plane hulls.* Since Ω is simply connected, there is a conformal map g from Ω onto $\mathbb{H}$. The standard normalization in this case is called the *thermodynamic normalization* and is given by condition $\lim_{z\to\infty}(g(z) - z) = 0$. The standard notation for this map is g_K and its inverse will always be denoted by f_K. Maps g_K are also called *mapping-out functions.* The existence and uniqueness of this normalization is given by the following lemma.

Lemma 2.15. *Let $\Omega = \mathbb{H} \setminus K$ where K is a half-plane hull. Then there is a unique conformal map $g_K : \Omega \to \mathbb{H}$ such that*

$$\lim_{z\to\infty}(g_K(z) - z) = 0.$$

Alternatively, its expansion at infinity is of the form

$$g_K(z) = z + b_1 z^{-1} + b_2 z^{-2} + \cdots .$$

The coefficient b_1 in front of z^{-1} has a geometrical meaning and, in some sense, it describes how large the set K is as seen from infinity. This coefficient is called the *half-plane capacity of* K and is denoted by $\mathrm{hcap}(K)$. We will discuss its properties in Section 3.5.

Proof of Lemma 2.15. Since Ω is simply connected, there is a conformal map g from Ω onto $\mathbb{H}$. By composing with an appropriate Möbius transformation, we can assume that $g(\infty) = \infty$.

Later on, we will show that g can be extended continuously onto the real part of the boundary of Ω; for now, we just assume this. By the Schwarz reflection principle, g can be extended to a univalent map on Ω', which is the complex plane without closure K and its symmetric image. This map has an isolated singularity at infinity and hence has a Laurent expansion at infinity. Since it is univalent in the neighbourhood of infinity, the series must be of the form

$$g(z) = a_1 z + a_0 + a_{-1} z^{-1} + a_{-2} z^{-2} + \cdots .$$

Since near infinity the real line is mapped onto the real line, the leading coefficient a_1 must be real, and since the points in the upper half-plane are mapped to the upper half-plane, it must be positive. This proves that $g(z)/a_1$ also maps Ω onto $\mathbb{H}$.

Next, we consider $g(z)/a_1 - z$; this map is real for large real z, which proves that a_0 must be real. Moreover, repeating the same argument, we prove by induction that *all* coefficients are real. Subtracting a_0/a_1, we find a conformal map $g_K : \Omega \to \mathbb{H}$ which has the following expansion at infinity:

$$g_K(z) = \frac{g(z) - a_0}{a_1} = z + b_1 z^{-1} + b_2 z^{-2} + \cdots .$$

The proof of uniqueness is very standard. Assuming that there are two such maps g_1 and g_2, we consider $g_1 \circ g_2^{-1}$, which is a conformal automorphism of $\mathbb{H}$ and hence a Möbius transformation. Direct computation shows that the only Möbius transformation of $\mathbb{H}$ with the thermodynamic normalization is the identity. □

Unbounded domains. There are many situations where one has to study simply connected domains of the form $\widehat{\mathbb{C}} \setminus K$ where K is a compact set. In this case, it might be more convenient to use the complement of the unit disc $\mathbb{D}_-$ as the uniformizing domain.

Theorem 2.16. *Let Ω be a simply connected domain in $\widehat{\mathbb{C}}$ such that infinity is an interior point. Then there is a unique conformal map f from $\mathbb{D}_-$ onto Ω such that $f(\infty) = \infty$ and the derivative at infinity is positive. Equivalently, the power series for f at infinity is of the form*

$$f(z) = a_1 z + a_0 + a_{-1} z^{-1} + \cdots ,$$

where $a_1 > 0$.

This theorem follows directly from the usual statement, but we will use this form several times later on, so we state it explicitly. It is also more convenient to state it in terms of the map from the uniformizing domain onto Ω, but this is a merely a cosmetic change.

Proof of Theorem 2.16. Without loss of generality, we can assume that $0 \in K$, otherwise we can shift K; this will change f by an additive constant. Consider the domain $\widetilde{\mathbb{C}} = \{z : 1/z \in \Omega\}$. Clearly, 0 is an interior point of $\widetilde{\mathbb{C}}$. By the Riemann theorem, there is a map $\tilde{f} : \mathbb{D} \to \widetilde{\mathbb{C}}$ such that $\tilde{f}(0) = 0$ and $\tilde{f}'(0) > 0$. Define $f(z) = 1/\tilde{f}(1/z)$; it is clear that this map is analytic and univalent in $\mathbb{D}_-$, and since $\tilde{f}$ has a simple zero at 0, the expansion of f at infinity is

$$f(z) = \frac{1}{\tilde{f}'(0)} z + a_0 + a_{-1} z^{-1} + \cdots .$$

Since $\tilde{f}'(0) > 0$, $f'(\infty) = 1/\tilde{f}'(0)$ is also positive. □

The constant $f'(\infty)$ is very important and very similar to the notion of conformal radius that we introduced before. In some sense, it measures how large the compact set K is. Sometimes, it is also convenient to work with $\log f'(\infty)$ which is called the *(logarithmic) capacity* of K and is denoted by $\mathrm{cap}(K)$. By the same argument as for the monotonicity of conformal radius, we can prove that capacity is monotone, namely, if $K_1 \subset K_2$, then $\mathrm{cap}(K_1) \leq \mathrm{cap}(K_2)$. We will discuss further properties of capacity in later chapters.

2.6. Constructive Proofs

In this section, we briefly discuss some constructive proofs of the Riemann mapping theorem. We present constructions, but do not give complete proofs. It is important to note that all constructive proofs give a series of increasingly better approximations to the Riemann map, but do not give the map itself. This should not be a surprise since there are very few domains for which the Riemann map can be written explicitly.

2.6.1. *Composition of elementary maps*

We assume that $\Omega \subset \mathbb{D}$ and that $0 \in \Omega$, otherwise we can repeat the explicit construction from the first step of the Riemann mapping theorem's proof. To construct the uniformization map, we use the last step from the proof.

We construct inductively a sequence of domains $\Omega = \Omega_1, \Omega_2, \Omega_3, \ldots$ and conformal maps f_n from Ω_n onto Ω_{n+1}. We show that $\Omega_n \to \mathbb{D}$ (in some sense) and the composition of f_n will converge to a conformal map from Ω onto $\mathbb{D}$.

Given Ω_n, define $r_n = \inf\{|z|, z \in \mathbb{D} \setminus \Omega_n\}$ and let w_n be some point in $\mathbb{D} \setminus \Omega_n$ with $|w_n| = r_n$. As in the proof of the Riemann mapping theorem, we define

$$\psi_n = \sqrt{\frac{z - w_n}{1 - \bar{w}_n z}}$$

and

$$f_n = \frac{|\psi'(0)|}{\psi'(0)} \frac{\psi(z) - \psi(0)}{1 - \psi(z)\overline{\psi(0)}}.$$

As before, we have that $f_n(0) = 0$ and $f_n'(0) = (1 + r_n)/2\sqrt{n} > 1$. Define $\Omega_{n+1} = f_n(\Omega_n)$. By construction, $F_n = f_n \circ f_{n-1} \circ \cdots \circ f_1$ is a univalent map from Ω onto $\Omega_{n+1} \subset \mathbb{D}$ with $F_n(0) = 0$ and

$$F_n'(0) = \prod_{i=1}^{n} f_i'(0) = \prod_{i=1}^{n} \frac{1 + r_i}{2\sqrt{r_i}}.$$

From the Schwarz lemma, we know that $|F_n'(0)|$ can be bounded by some constant which depends on Ω and z_0 only, but not on Ω_n. Since the product is increasing and bounded, it must converge. In particular, this implies that $f_n'(0) \to 1$ or, equivalently, $r_n \to 1$ as $n \to \infty$. This means that Ω_n is squeezed between $r_n\mathbb{D}$ and $\mathbb{D}$ and hence converges to $\mathbb{D}$ (in the Hausdorff topology). It is also possible to show that the sequence of maps F_n converges uniformly on all compact subsets of Ω and that the limiting function is a univalent map from Ω onto $\mathbb{D}$. To prove it rigorously, we need the Carathéodory convergence theorem (Theorem 3.16), which we will prove in Section 3.3.

This construction follows the same idea that the uniformizing map should maximize the derivative at the point which should be mapped to the origin but, instead of an abstract compactness argument, we use an explicit construction. Another advantage of this approach is that all functions f_n are elementary and easy to compute: they are compositions of Möbius transformations and square root functions. From a purely practical point of view, it might be difficult to compute r_n, but it is easy to see that we do not really need r_n to be optimal; we just need it to be comparable to the optimal.

2.6.2. *The Christoffel–Schwarz formula*

The next method uses domain approximations. The main idea is to approximate a domain by polygonal domains, and for a polygonal domain there is a nice expression for a conformal map from the unit disc onto these domains, given by the Christoffel–Schwarz formula.[6] Detailed discussion of Christoffel–Schwarz maps can be found in Driscoll and Trefethen (2002). Here, we just provide a brief description.

[6]Most authors prefer to call it the Schwarz–Christoffel formula.

Theorem 2.17 (Christoffel–Schwarz). *Let Ω be a polygonal domain with n vertices where angles between adjacent edges are equal to $\pi\alpha_k$. Then there is a conformal map from $\mathbb{D}$ onto Ω which has the form*

$$F(w) = C\int_0^w \prod_{k=1}^{n}(w - w_k)^{-\beta_k}\mathrm{d}w + C',$$

where $\beta_k = 1 - \alpha_k$, w_k are some points on the unit circle, and C and C' are complex-valued constants.

There is an alternative version for a map from the upper half-plane. In this case, the mapping is given by

$$F(w) = C\int_0^w \prod_{k=1}^{n-1}(w - x_k)^{-\beta_k}\mathrm{d}w + C', \tag{2.1}$$

where x_i are real numbers. This function maps $x_1 \ldots x_{n-1}$ onto $n-1$ vertices, and the last vertex is the image of ∞.

It is possible to explain the origin of a somewhat mysterious product in (2.1) without going into a detailed analysis of these maps. Let us consider the image of the real line. It is easy to see that between two consecutive points x_{k-1} and x_k, the argument of the product is constant, and hence its image is a straight interval. When w moves from this interval to the next $[x_k, x_{k+1}]$, the arguments of most of the factors stay the same; the only difference is that the argument of $(w - x_k)^{-\beta_k}$ changes from $-\pi\beta_k$ to 0. This means that the angle between the images of $[x_{k-1}, x_k]$ and $[x_k, x_{k+1}]$ is $\pi\beta_k$ (see Figure 2.2). This means that at least the image of the real line is a polygonal curve with the prescribed angles. Checking when this curve is simple is less trivial.

The main disadvantage of this formula is that it is not as explicit as it looks: in practice, it is very difficult to compute the points w_k. Even when the points w_k are known, the map is given by an integral which has integrable singularities, which make it not very amenable to straightforward computations. Banjai and Trefethen (2003) invented a fast algorithm for computing the Christoffel–Schwarz maps.

There are very few cases when the Christoffel–Schwarz function can be computed explicitly, even in terms of special functions. One of the main problems is the difficulty of computing the parameters w_i in the disc version, or x_i in the half-plane version. In the following, we work out a few examples where this can be done.

Christoffel–Schwarz map for a triangle. To compute the Christoffel–Schwarz map from the upper half-plane to a triangle, we have to choose the

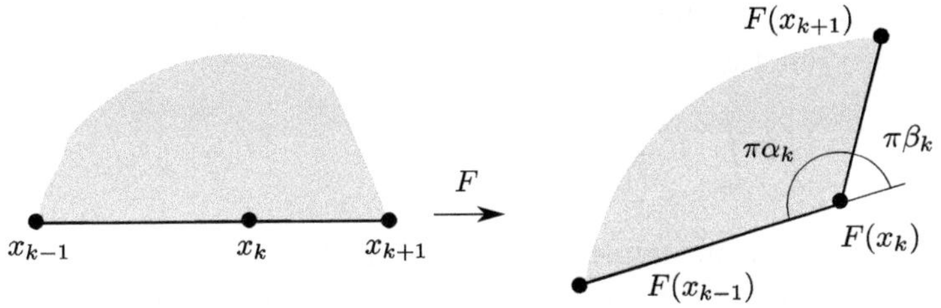

Fig. 2.2. Christoffel–Schwarz function changes the argument at x_k by $\pi\beta_k$.

parameters x_i. Fortunately, for a triangle, the choice is irrelevant. Indeed, for any x_1 and x_2 and any β_1 and β_2, the function defined by (2.1) maps the upper half-plane onto a triangle with angles $\pi(1-\beta_1)$, $\pi(1-\beta_2)$, and $\pi(\beta_1+\beta_2-1)$. Since all triangles with the same angles are similar, one can always choose constants C and C' in such a way that the image is the prescribed triangle.

There are two cases when the formulas are particularly simple: a) the equilateral triangle where $\beta_i = 2/3$; and b) the semi-infinite strip which can be thought of as a triangle with angles $\pi/2$, $\pi/2$, and 0 (equivalently $\beta_1 = \beta_2 = 1/2$, $\beta_3 = 1$).

Exercise 2.10. Find explicit formulas (that might involve special functions) for conformal maps between Ω and one of the standard uniformizing domains where Ω is

(1) a semi-infinite strip $S = \{z : -\pi/2 < \operatorname{Re} z < \pi/2,\ \operatorname{Im} z > 0\}$,
(2) an equilateral triangle.

The Christoffel–Schwarz formula for a rectangle. In the case of a rectangle, we have four vertices x_i, three of which could be chosen arbitrarily. One of the standard choices is to select them to be 0, $x_0 \in (0,1)$, 1, and ∞. The Christoffel–Schwarz formula could be written in this setup, but it is not optimal since it does not use the existing symmetries. Instead, we choose them to be ± 1 and $\pm 1/k$, where $k \in (0,1)$ is a parameter which will eventually define the shape of the rectangle. Technically, this corresponds to a pentagon where $\alpha_5 = 1$, so the fifth vertex could be removed. All angles are $\pi/2$ and $\beta_i = 1/2$. In this case, our integral can be written (up to a constant factor)

$$F(w) = \int_0^w \frac{dw}{\sqrt{(1-w^2)(1/k^2-w^2)}},$$

where arguments are chosen in such a way that the integrand is positive on $(-1,1)$. This function can be written as

$$F(w) = k\int_0^w \frac{dw}{\sqrt{(1-w^2)(1-k^2w^2)}} = kF(\arcsin(w); k^2),$$

where $F(w;k)$ is the incomplete elliptic integral of the first kind. To simplify notation, we redefine F by dropping the factor of k in front of the integral.

By the Christoffel–Schwarz formula, $F(w)$ maps $\mathbb{H}$ onto a rectangle with the vertices $F(\pm 1)$ and $F(\pm 1/k)$. First, we observe that this map could be continuously extended to the real line and that it maps $[-1,1]$ onto $[-K,K]$, where

$$K = K(k) = \int_0^1 \frac{dw}{\sqrt{(1-w^2)(1-k^2w^2)}}$$

is the complete elliptic integral of the first kind. To compute the images of two other points, we note that for $1 < w < 1/k$, we have

$$F(w) = \int_0^1 \frac{dw}{\sqrt{(1-w^2)(1-k^2w^2)}} + i\int_1^w \frac{dw}{\sqrt{(w^2-1)(1-k^2w^2)}},$$

so $F(1/k) = K + iK'$, where

$$K' = K'(k) = \int_1^{1/k} \frac{dw}{\sqrt{(w^2-1)(1-k^2w^2)}}$$

is the complementary complete elliptic integral of the first kind. It is a standard fact that $K'(k) = K(k')$ where $k' = \sqrt{1-k^2}$ is the complementary parameter. In other words, $F(w)$ maps $\mathbb{H}$ onto a rectangle with vertices $\pm K$ and $\pm K + iK'$ (Figure 2.3). The ratio of the horizontal side to the vertical one is

$$\lambda(k) = \frac{2K(k)}{K'(k)}.$$

Exercise 2.11. Let R and R' be two rectangles and let $\lambda, \lambda' \geq 1$ be the ratio of side lengths of R and R'. Assume that there is a conformal map $f : R \to R'$ which is continuous up to the boundary and maps vertices to vertices. Show that $\lambda = \lambda'$.

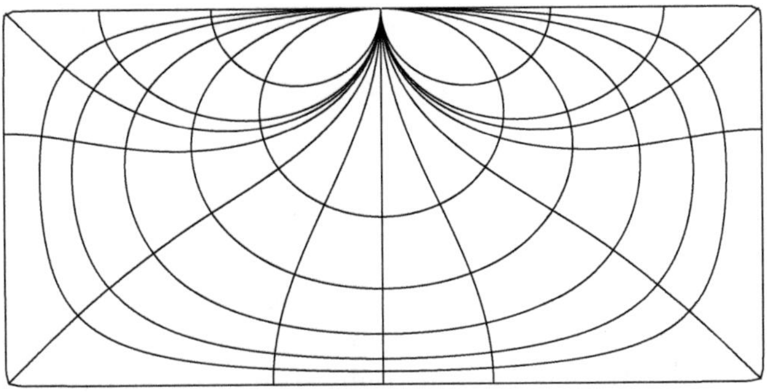

Fig. 2.3. Image of $\mathbb{H}$ with $k = 1/\sqrt{2}$. The lines inside are the images of horizontal and vertical lines.

2.7. Boundary Correspondence

In the previous sections, we discussed the existence of univalent maps from general domains onto simple uniformizing domains. These maps are analytic inside the corresponding domains, but *a priori* we have no information about their boundary behaviour. In this section, we investigate this question and obtain a simple geometrical answer.

First, we note that by means of elementary maps that are obviously continuous on the boundary, we can map any domain onto a bounded domain. This means that, without loss of generality, we can always assume that all domains in this section are bounded.

Next, we make a very simple observation which is purely topological and does not use analyticity: boundaries are mapped onto each other. The precise statement in the general case is given by the following proposition.

Proposition 2.18. *Let f be a univalent map from Ω onto Ω' and let $z_n \in \Omega$ be a sequence which tends to the boundary of Ω, which means that all accumulation points are on the boundary of Ω, equivalently,* $\operatorname{dist}(z_n, \partial\Omega) \to 0$. *Then $f(z_n)$ tends to the boundary of Ω'. Alternatively, one can assume that f naturally extends to a continuous function between one-point compactifications of Ω and Ω'.*

Proof. It is easy to see that the condition that z_n tends to the boundary is equivalent to the fact that for every compact $K \subset \Omega$, there is N such that z_n is outside of K for $n > N$. Let K' be a compact set in Ω'; then, by continuity, $K = f^{-1}(K')$ is also a compact set. Since z_n will eventually leave K, $f(z_n)$ will leave K'. □

2.7.1. *Accessible points*

Proposition 2.18 shows that the boundary as a whole set is mapped to the boundary, but it does not tell us anything about the continuity at the individual points on the boundary. The boundary behaviour of analytic functions is a rich and well-developed subject, but most of it is beyond the scope of this book. Here, we will use only some rather elementary considerations, which are surprisingly sufficient since we work with a rather small class of univalent functions. We start by considering boundary behaviour near 'regular' boundary points.

Definition 2.19. An *accessible boundary point* ζ of a domain Ω is an equivalence class of continuous curves $\gamma : [0,1] \to \bar{\Omega}$ which join a given point $\tilde{\zeta} \in \partial\Omega$ with an arbitrary interior point. We assume that γ lies completely inside Ω except $\gamma(1) = \tilde{\zeta}$. Two curves are equivalent if for an arbitrary neighbourhood U of $\tilde{\zeta}$, parts of the curves that are inside of $\Omega \cap U$ can be joined by a continuous curve inside $\Omega \cap U$.

If the boundary of Ω is nice, say, a Jordan curve, then each boundary point corresponds to exactly one accessible point. In this case, we identify them. It could also be that one boundary point corresponds to more than one accessible point, for example, see Figure 2.4. In Figure 2.4(a), ζ is a point on a boundary such that $B(\zeta, r) \cap \Omega$ has only one component for sufficiently small r. This point must correspond to one accessible point. In Figure 2.4(b), ζ is a point on a slit and $B(\zeta, r) \cap \Omega$ has two components, each of which gives rise to an accessible point. The last example in Figure 2.4(c) is more involved. Let $\zeta = 0$ and for each dyadic direction, $\theta = 2\pi k/2^n$ (we assume that $n \in \mathbb{N}$ and that $k > 0$ is odd), we remove an interval $[0, e^{i\theta}/2^n]$. For each irrational (mod 2π) angle θ, we can consider $\gamma_\theta(t) = (1-t)e^{i\theta}$. It

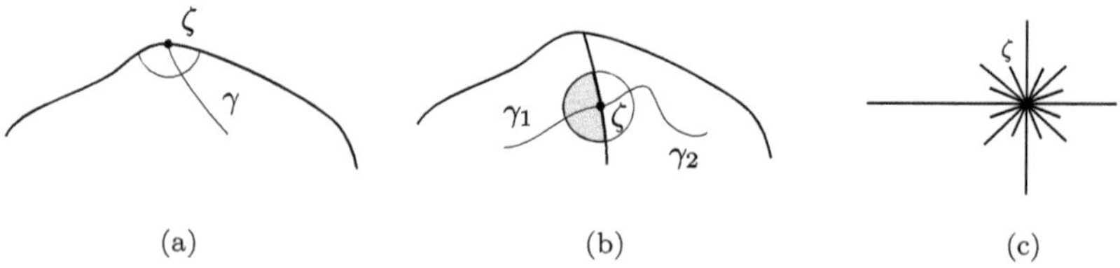

Fig. 2.4. A single boundary point could correspond to one (a), two (b), or even uncountably many (c) accessible points.

is not very difficult to see that each γ_θ defines an accessible point and that all these accessible points are different.

Note that accessible points that correspond to different boundary points are always different, but the same boundary point could correspond to several accessible points. If accessible points are different then, for sufficiently small r_0, there are disjoint components of $B(\zeta_i, r_0) \cap \Omega$ such that the tails of the curves defining accessible points lie in the corresponding components. We denote these components $U(\zeta, r_0)$. Sometimes, it is beneficial to identify an accessible point with the corresponding component $U(\zeta, r)$ for small r. One should be careful doing so: for every two accessible points, there are neighbourhoods separating them, but if there are infinitely many accessible points corresponding to the same boundary point, then it could be that there is no single r_0 which allows one to separate a particular accessible point from all others. An example is given in Figure 2.4(c).

It could also be that there are no continuous curves γ approaching a boundary point; in this case, the boundary point is not accessible at all; see Figures 2.5 and 2.6.

Theorem 2.20. *Let Ω be a simply connected bounded domain in the plane and let f be a univalent map from Ω onto $\mathbb{D}$. Then, for every accessible point ζ, the map f can be continuously extended to ζ in such a way that $|f(\zeta)| = 1$. Moreover, for distinct accessible points, their images are distinct.*

There are several ways to prove this theorem. One of the standard modern approaches is to consider the inverse function and use some powerful results of the existence of the radial limits for functions from the Hardy

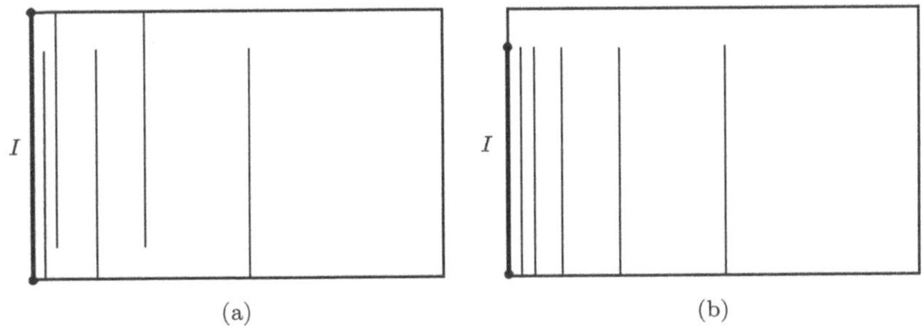

Fig. 2.5. In both examples, we remove from a rectangle a sequence of intervals that are getting closer and closer to the right boundary. On the left, all points of the closed interval I are not accessible. On the right, all points of the half-open interval I are not accessible, except the top end which is accessible.

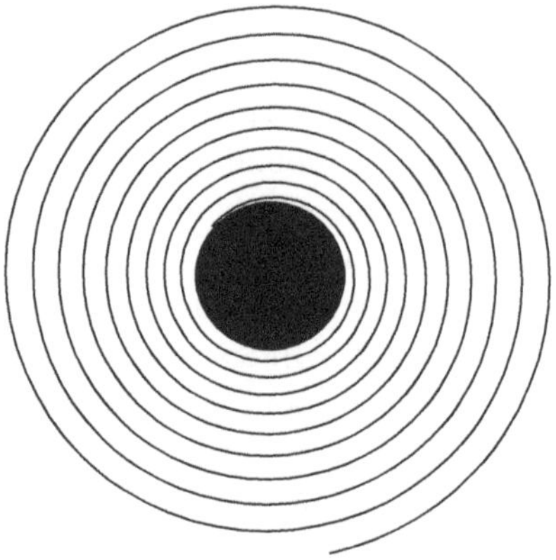

Fig. 2.6. The central disc is approached by a slow spiral which goes infinitely many times around the disc. The disc cannot be approached by a rectifiable curve; hence all points on its boundary are not accessible. Each point on the spiral corresponds to two accessible points.

class H^∞. Here, we prefer to give a rather elementary geometrical proof based on a technical lemma which is due to Koebe.

Lemma 2.21 (Koebe). *Let z_n and z'_n be two sequences in the unit disc $\mathbb{D}$ converging to two distinct points ζ and ζ' on the boundary of the unit disc. Let γ_n be Jordan arcs connecting z_n and z'_n inside $\mathbb{D}$ but outside some fixed neighbourhood of the origin. Finally, assume that a function f is analytic and bounded in $\mathbb{D}$ and that f converges uniformly to 0 on γ_n, that is, the sequence $\varepsilon_n = \sup_{\gamma_n} |f|$ converges to 0. Then f is identically equal to 0 in $\mathbb{D}$.*

Proof. Let us assume that f is not identically zero. Without loss of generality, we assume that $f(0) \neq 0$, otherwise f has zero of the finite order n and we can replace f by $f(z)/z^n$ which satisfies all assumptions of the lemma.

For sufficiently large m, there is a sector S of angle $2\pi/m$ such that the radii towards ζ and ζ' lie outside of this sector and infinitely many γ_n cross this sector. We discard all other curves as well as their endpoints. Abusing notation, we call the remaining curves γ_n. By rotating the unit disc, that is, by considering $f(e^{i\alpha}z)$ instead of $f(z)$, we can assume that the positive real line is the bisectrix of S.

For each curve γ_n, we can find its part γ'_n which is also a simple curve that crosses the upper half of S: its endpoints lie on two different radii

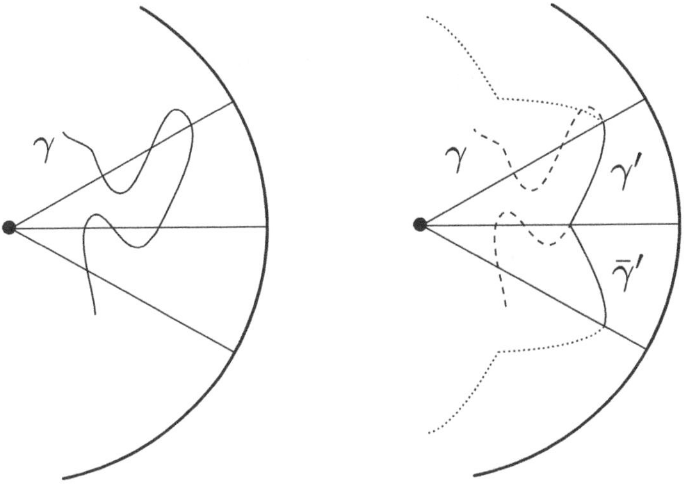

Fig. 2.7. There is a part of the original curve γ which crosses the upper half of the sector. Rotations of this part γ', together with its symmetric image $\bar{\gamma}'$, form a simple closed curve around the origin.

bounding the upper half of S, and no other point lies on these radii. Define $\bar{\gamma}_n'$ to be symmetric to γ_n' about the real axis (see Figure 2.7).

By the reflection principle, the function $\bar{f}(\bar{z})$ is also analytic in $\mathbb{D}$ and it is bounded by ε_n on $\bar{\gamma}_n'$. This means that the function $\phi(z) = f(z)\bar{f}(\bar{z})$ is analytic and bounded on the union of γ_n' and $\bar{\gamma}_n'$ by $M\varepsilon_n$, where $M = \sup_{\mathbb{D}} |f|$.

Let F be the product of rotations of ϕ by $2\pi/m$, namely

$$F(z) = \phi(z)\phi(e^{2\pi i/m}z)\dots\phi(e^{2\pi i(m-1)/m}z).$$

This function is analytic in $\mathbb{D}$ and bounded by $\varepsilon_n M^{2m-1}$ on a closed curve formed by the union of rotations of γ_n' and $\bar{\gamma}_n'$. By the maximum modulus principle, this implies that $|f(0)|^{2m} = |F(0)| \le \varepsilon_n M^{2m-1}$ for every n. Since $\varepsilon_n \to 0$, this implies that $f(0) = 0$, which contradicts our initial assumption. □

The following lemma is known as the *length–area argument* or Wolff's lemma. Despite its simplicity, it is a very powerful result that is extensively used in geometric function theory.

Lemma 2.22 (Length–area). *Let $z_0 \in \mathbb{C}$ be a fixed point and $\Omega \subset \mathbb{C}$ be any domain. For $r > 0$, we define $\tilde{S}_r$ to be $\{z \in \Omega : |z - z_0| = r\}$. Let f be a function univalent in Ω such that $f(\Omega)$ has finite area. We denote the image of $\tilde{S}_r$ under f by S_r, and the length of S_r by $l(S_r)$. Then for any*

$r_1 < r_2$, we have

$$\liminf_{r_1<r<r_2} l(S_r) \leq \frac{\sqrt{2\pi \text{Area}(f(\Omega))}}{\sqrt{\log(r_2/r_1)}}. \tag{2.2}$$

In particular, there is a sequence $r_n \to 0$ such that

$$l(S_{r_n}) \to 0, \quad n \to \infty.$$

Proof. Without loss of generality, we can assume that $z_0 = 0$, otherwise we can shift everything by z_0.

We can write the length of S_r as

$$l^2(S_r) = \left(\int_{\tilde{S}_r} |f'(w)|\mathrm{d}w\right)^2 = \left(\int |f'(re^{i\theta})|r\mathrm{d}\theta\right)^2,$$

where, in the last integral, we integrate over all θ such that $re^{i\theta} \in \Omega$. Applying the Cauchy–Schwarz inequality, we obtain

$$l^2(S_r) \leq \left(\int |f'(re^{i\theta})|^2 r\mathrm{d}\theta\right)\left(\int r\mathrm{d}\theta\right) \leq 2\pi r \int |f'(re^{i\theta})|^2 r\mathrm{d}\theta.$$

Dividing by r and integrating with respect to r, we have

$$\begin{aligned}\left(\liminf_{r_1<r<r_2} l^2(S_r)\right)\log(r_2/r_1) &= \left(\liminf_{r_1<r<r_2} l(S_r)\right)^2 \int_{r_1}^{r_2} \frac{\mathrm{d}r}{r} \leq \int_{r_1}^{r_2} \frac{l(S_r)^2}{r}\mathrm{d}r \\ &\leq 2\pi \int_{r_1}^{r_2} \int |f'(re^{i\theta})|^2 r\mathrm{d}\theta\mathrm{d}r \leq 2\pi \int_\Omega |f'(z)|^2 \\ &= 2\pi\,\text{Area}(f(\Omega)).\end{aligned}$$

This proves the first part. To prove the second claim, we fix some r_2 and send r_1 to zero. By (2.2),

$$\liminf_{0<r<r_2} l(S_r) = 0,$$

which immediately implies that there is a sequence along which the length tends to zero. □

Proof of Theorem 2.20. Let $\gamma(t)$ be a curve defining the accessible point ζ. We want to show that $\tilde{\gamma}(t) = f(\gamma(t))$ converges to a point on the unit circle as $t \to 1$. Later on, we will show that this limit is independent of a particular choice of γ.

Let us assume the contrary; then $\tilde{\gamma}$ contains a sequence of arcs with endpoints converging to two distinct boundary points (see Figure 2.8). Moreover, these arcs converge to the boundary and hence stay away from the

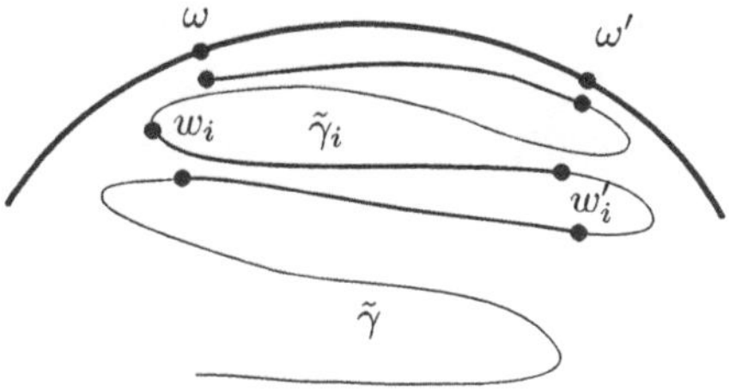

Fig. 2.8. Thick parts of the curve γ form arcs whose endpoints converge to two distinct points ω and ω'.

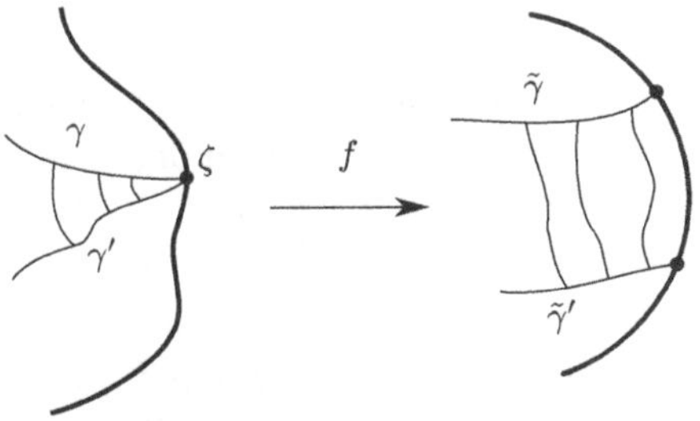

Fig. 2.9. Images of arcs connecting curves γ and γ' form a sequence of arcs in $\mathbb{D}$ whose endpoints converge to two distinct points on the unit circle.

origin. The inverse function $g = f^{-1}$ converges uniformly to ζ on these arcs. Applying the Koebe lemma (Lemma 2.21) to $g - \zeta$, we see that g must be identically equal to ζ, which is obviously impossible. This proves that as we move along $\tilde{\gamma}$, we must approach a definite point on the unit circle. In other words, the limit of $f(\gamma(t))$ must exist and we define $f(\zeta)$ to be this limit.

Next, we have to show that this definition is consistent, that is, independent of our choice of γ. Let $\gamma'(t)$ be another curve describing the same accessible point ζ. As before, we know that $\tilde{\gamma}' = f(\gamma')$ approaches a single point on the unit circle. We assume that $\tilde{\gamma}$ and $\tilde{\gamma}'$ approach two distinct points. By the definition of accessible point, curves γ and γ' can be connected by a Jordan arc within any neighbourhood of ζ. As these neighbourhoods contract to ζ, their images become arcs whose endpoints converge to two distinct points on the unit circle (see Figure 2.9). On these arcs, g converges uniformly to ζ and, as before, this implies that g is constant.

Let ζ and ζ' be two different accessible points corresponding to the same boundary point which we denote by ξ (the case when accessible points correspond to different boundary points is trivial and omitted). Let γ and

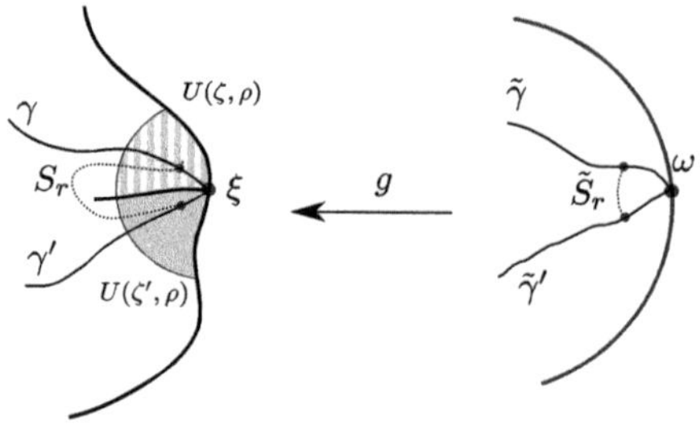

Fig. 2.10. When ζ and ζ' are two different accessible points, the length of a curve connecting two points on γ and γ' close to ξ, respectively, cannot be too small.

γ' be the corresponding curves, and $U(\zeta,\rho)$ and $U(\zeta',\rho)$ be the disjoint components of $B(\xi,\rho)\cap\Omega$ as in the definition of accessible points. We know that $f(z)$ approaches definite points on the unit circle as z approaches ξ along γ or γ'. We assume that they approach the same point $\omega\in\partial\mathbb{D}$ and show that it leads to a contradiction (see Figure 2.10).

As before, we denote $f(\gamma)$ and $f(\gamma')$ by $\tilde{\gamma}$ and $\tilde{\gamma}'$. They are two continuous curves in $\mathbb{D}$ that converge to the same $\omega\in\partial\mathbb{D}$. For sufficiently small r_0 and all $r<r_0$, there are arcs $\tilde{S}_r$ of the circle $|w-\omega|=r$ such that they lie in $\mathbb{D}$ and they have one endpoint in $\tilde{\gamma}$ and the other in $\tilde{\gamma}'$. We denote $g(\tilde{S}_r)$, a continuous curve in Ω connecting γ and γ', by S_r. Since endpoints of $\tilde{S}_r$ converge to ω along $\tilde{\gamma}$ and $\tilde{\gamma}'$ as $r\to 0$, the endpoints of S_r converge to ξ.

Applying the length–area lemma (Lemma 2.22) to g, we have that $l(S_r)$ cannot be bounded away from zero. This immediately implies that both accessible points correspond to the same boundary point. If two accessible points are different then, for some ρ, the tails of γ and γ' are in two different components of $\Omega\cap B(\xi,\rho)$. This means that the length of every curve connecting two points on γ and γ' that are in $B(\xi,\rho/2)$ must be at least ρ. This contradicts our previous result that there are such curves with arbitrarily small length. Hence, ζ and ζ' must be the same accessible point. □

Since all points on a Jordan curve correspond to exactly one accessible point, Theorem 2.20 immediately implies the following result, known as the Carathéodory theorem.

Theorem 2.23 (Carathéodory). *Let Ω be a simply connected domain bounded by a closed Jordan curve Γ and let f be a conformal map from Ω onto $\mathbb{D}$. Then f could be continuously extended to a bijection from Γ onto the unit circle.*

Proof. Existence of the extension and the fact that it is a bijection follows directly from Theorem 2.20. The continuity follows from monotonicity of the argument. The details are left to the reader. □

There is an alternative formulation in terms of the inverse function g.

Theorem 2.24 (Carathéodory). *Let Ω be a simply connected domain and let g be a conformal map from $\mathbb{D}$ onto Ω. Then g continuously extends to a homeomorphism between $\bar{\mathbb{D}}$ and $\bar{\Omega}$ if and only if Ω is a Jordan domain, that is, its boundary is a Jordan curve.*

The 'if' part is exactly equivalent to the previous formulation and the 'only if' part is trivial, since the boundary of Ω is a continuous injective image of the unit circle.

It is not surprising that for analytic boundaries, the result is even stronger (but the proof is beyond the scope of this book).

Theorem 2.25. *Let Ω be a domain bounded by an analytic Jordan curve. Then a conformal map f from Ω onto $\mathbb{D}$ can be extended to a function analytic on the boundary.*

If we do not require the map to be injective on the unit circle, just continuous, then we have the following result.

Theorem 2.26 (Carathéodory). *Let Ω be a simply connected domain and let g be a conformal map from $\mathbb{D}$ onto Ω. Then g continuously extends to the boundary of $\mathbb{D}$ if and only if the boundary of Ω is locally connected, that is, for every point $z \in \partial\Omega$, there exists r such that $\partial\Omega \cap B(z,r)$ is connected.*

Remark 2.27. Local connectivity is closely connected with accessibility. In particular, boundaries of all domains in Figure 2.4 are locally connected and those in Figure 2.5 are not. There is an equivalent formulation of local connectivity. In the case of the boundaries of two-dimensional domains, local connectivity is equivalent to the statement that the boundary of Ω is a curve, that is, a continuous (but, of course, not necessarily injective) image of the unit circle. This result is known as the Hahn–Mazurkiewicz theorem, which can be found in many topology books (see, for example, Hocking and Young (1988, Theorem 3-30)). In our context, when the map $g : \mathbb{D} \to \Omega$ can be continuously extended to the boundary, we could parametrize $\partial\Omega$ by $g(e^{i\theta})$.

Surprisingly, the inverse boundary correspondence holds.

Theorem 2.28. *Let f be analytic in Ω and continuous up to the boundary. Let us assume that the boundary of Ω is a positively oriented Jordan curve Γ. If f is a continuous orientation preserving bijection from Γ onto another Jordan curve Γ', then f is a univalent map from Ω onto the domain Ω' bounded by Γ'.*

Proof. Let w_0 be any point in Ω'. Since f maps Γ onto Γ', we have that $f \neq w_0$ on Γ. By continuity, there is a neighbourhood $U \subset \Omega$ of Γ where $f \neq w_0$ as well.

For any closed curve $\gamma \subset \Omega$, we can consider the quantity

$$\frac{1}{2\pi}\Delta_\gamma \arg(f(z) - w_0)$$

the normalized increment of the argument along γ. It is easy to see that when we continuously deform γ, this quantity changes continuously as long as the $f(\gamma)$ does not cross w_0. Since this quantity is integer-valued for any closed curve, it must be constant for all curves that are continuous deformations of each other inside U.

By the theorem's assumptions,

$$\frac{1}{2\pi}\Delta_\Gamma \arg(f(z) - w_0) = \frac{1}{2\pi}\Delta_{\Gamma'} \arg(w - w_0) = 1.$$

Let $\gamma \subset \Omega$ be a simple curve homotopic to Γ inside U. The above argument implies that the same is true for γ. Let D be the domain bounded by γ. For this domain, we can apply the argument principle and obtain that the equation $f = w_0$ has exactly one solution inside D. But $\Omega \setminus D \subset U$ and by construction, $f \neq w_0$ here. This proves that there is a unique point $z_0 \in \Omega$ such that $f(z_0) = w_0$.

By the same argument, $f \neq w$ for every w in the interior of the complement of Ω'. Finally, no point of Ω is mapped onto a point of Γ', otherwise its neighbourhood would be mapped onto a neighbourhood of a point on the boundary of Ω' and there will be points outside, which contradicts the above argument. □

Note that in the previous theorem, we can assume that Ω is a domain with Jordan boundary in $\widehat{\mathbb{C}}$. But the domain Ω' should be bounded, as can be seen from the following simple example.

Example 2.29. Let $\Omega = \Omega'$ be the upper half-plane, the boundary $\Gamma = \Gamma' = \mathbb{R}$. Function $f(z) = z^3$ is a continuous bijection from Γ onto Γ', but it does not map Ω onto Ω'.

2.7.2. *Prime ends*

Considering simple examples of slit domains where the uniformizing maps are known explicitly, we can see that these maps are not continuous if one uses the ordinary Euclidean topology. Maps obviously behave differently on different sides of the slits (see, for example, Exercise 2.8 and Figure 2.19). From the internal geometry point of view, two points on the different sides of the slit are far away. The notion of an accessible point formalizes this intuition and allows one to treat the two sides of a slit as two different sets. This allows us to study boundary behaviour for all domains with a relatively simple boundary. To complete the study of the boundary correspondence, we have to study what happens at non-accessible points. For this, we need the notion of prime ends introduced by Carathéodory (1913). In this section, we assume that all domains are bounded. This is not too important but simplifies some computations and allows us to avoid using the spherical metric.

Definition 2.30. A *cross-section* or a *cross-cut* in a simply connected domain Ω is a Jordan arc $\gamma : (0,1) \to \Omega$ such that the limits $\gamma(t)$ as t approaches 0 or 1 exist and lie on the boundary of Ω. Curve γ separates Ω into two connected domains. We assume that the boundaries of both domains contain boundary points of Ω other than the endpoints of γ.

It is easy to see that the endpoints of a cross-cut must be different accessible points.

Definition 2.31. A *null chain* or *zero chain* is a sequence of cross-cuts γ_n such that closures of γ_n and γ_{n+1} are disjoint for all n, the diameter of γ_n tends to zero, and any Jordan curve in Ω connecting a point on γ_n with a given point $z_0 \in \Omega$ must intersect all γ_m for $m < n$.

Definition 2.32. We say that two chains γ_n and γ'_n are equivalent if for every n the arc γ_n separates almost all γ'_m from γ_{n-1} and γ'_n separates almost all γ_m from γ'_{n-1}. A *prime end* is an equivalence class of chains. The collection of all prime ends of Ω is denoted by $P(\Omega)$.

Figure 2.11 illustrates these definitions.

There is an alternative way to define the equivalence of the chains. Let D_n be the connected component of $\Omega \setminus \gamma_n$ which does not contain γ_{n-1}. This component is called the *inside* of γ_n. It contains all γ_m with $m > n$. It is easy to see that $D_n \supset D_{n+1}$. Let D_n and D'_n be two collections of

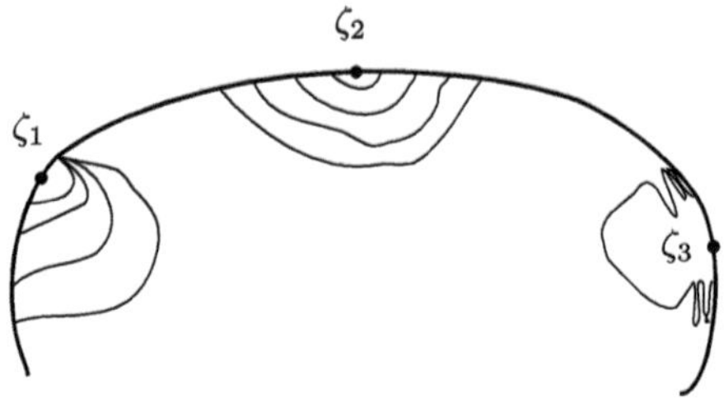

Fig. 2.11. Curves near ζ_1 do not form a chain. A curve near point ζ_3 is not a cross-cut since it is not continuous at the endpoints. Curves near ζ_2 form a chain that defines a prime end which corresponds to an accessible point.

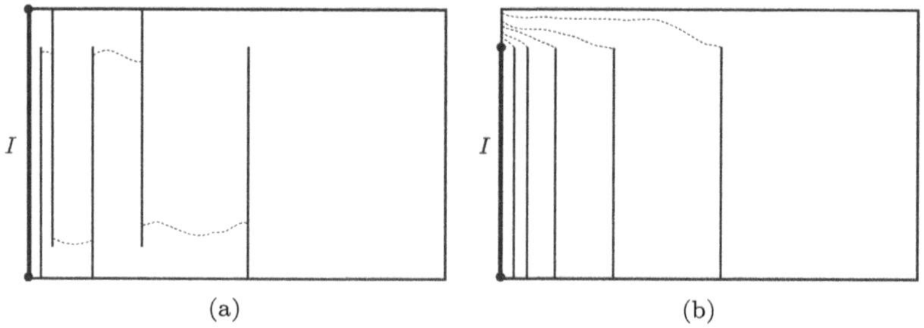

Fig. 2.12. In both examples, the interval I is the impression of the corresponding prime end. Dashed lines form null chains.

subdomains corresponding to two chains. Then the chains are equivalent if and only if each domain from one collection contains all but finitely many domains from the other collection. Using this notion, we can define a prime end by the condition that diameters of $f(D_n)$ tend to zero instead of the diameters of γ_n. This might be useful sometimes, but this approach has an obvious downside: it is not a purely geometric definition as it depends on the analysis of f.

Definition 2.33. Let p be a prime end and $(\gamma_n) \in p$ be one of the null-chains in p. The *impression* of a prime end p is defined as $\bigcap_n \bar{D}_n$ where $\bar{D}_n$ is the closure of the inside of γ_n. We denote it by $I(p)$.

Figure 2.12 shows examples of prime ends with non-trivial impressions.

Exercise 2.12. Prove that the impression of a prime end is well defined, that is, it does not depend on the choice of the null chain.

Another simple observation is that each accessible point can be associated with a prime end. Indeed, let us consider an accessible point ζ which is defined by a curve γ. We can define γ_n to be arcs of the circles $|\zeta - z| = 1/n$ that intersect with γ. These arcs form a chain and the impression of the corresponding prime end is ζ. Clearly, for different accessible points, these prime ends are different.

Now we can formulate the most general result of the boundary correspondence.

Theorem 2.34 (Carathéodory). *Let Ω be a simply connected domain and let f be a conformal map from $\mathbb{D}$ onto Ω. Then f can be extended to a bijection $\partial\mathbb{D} \to P(\Omega)$. Moreover, the extension can be defined such that for every ω and a null chain (c_n) representing $f(\omega)$ its preimage $(f^{-1}(c_n))$ is a null chain with impression $\{\omega\}$. In other words, this is the natural extension.*

In fact, a stronger result holds. It is possible to define a natural topology on $\Omega \cup P(\Omega)$ and the extension given by the above theorem is a homeomorphism of the closed disc onto $\Omega \cup P(\Omega)$. The proof can be found in Conway (1995, Theorem 14.3.4).

Proof of Theorem 2.34. First, we define the extension. Let us consider the circular cross-cuts $\tilde{S}_r = \mathbb{D} \cap \{|z - \omega| = r\}$. By Lemma 2.22, there is a sequence $r_n \searrow 0$ such that lengths of the images $S_{r_n} = f(\tilde{S}_{r_n})$ tend to 0. In particular, the length of S_{r_n} is finite; this implies that S_{r_n} is a cross-cut. The sequence of these cross-cuts almost forms a null chain. The only problem is that it might be that different cross-cuts have the same endpoints, and hence they are not separated. To avoid this, we modify cross-cuts a little bit.

Let A_n be the disjoint annular neighbourhoods of $\tilde{S}_{r_n}$. For example, we can take

$$A_n = \{z \in \mathbb{D} : r_n - (r_n - r_{n+1})/3 < |z - \omega| < r_n + (r_{n-1} - r_n)/3\}.$$

We will define a null chain inductively. We start with $\gamma_1 = \tilde{S}_{r_1}$. Let us assume that γ_k for $k < n$ are already defined. If the endpoints of S_{r_n} are distinct from the endpoints of $f(\gamma_k)$ for all $k < n$, then we define $\gamma_n = \tilde{S}_{r_n}$. Otherwise, we can consider the family of circular arcs around this joint endpoint. Applying the length–area lemma to f^{-1}, we can find circular arcs with arbitrarily small preimages. In particular, we could find an arc such that its endpoints are distinct from the endpoints of all $f(\gamma_k)$ and its preimage is in A_n. Replacing a part of S_{r_n} by a part of such an arc, we obtain a cross-cut in Ω which is separated from all previous cross-cuts, such

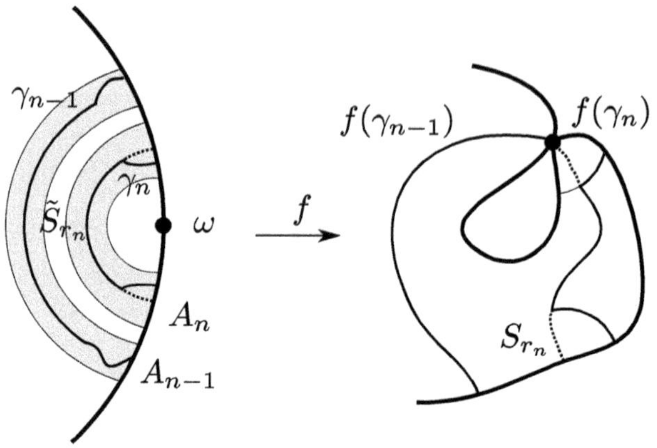

Fig. 2.13. Cross-cuts S_{r_n} in Ω could be modified near the endpoints so that all modifications have different endpoints and their preimages γ_n are still in the corresponding annuli A_n.

that its length is at most twice the length of S_{r_n}. We define γ_n to be its preimage (see Figure 2.13).

By construction, both (γ_n) and $(f(\gamma_n))$ are null chains. We define $f(\omega)$ to be the prime end defined by $(f(\gamma_n))$. It is easy to see that the prime end $(f(\gamma_n))$ is independent of the choice of r_n.

To show that the extension is one-to-one, we consider two distinct points ω and ω' on $\partial\mathbb{D}$. Let γ_n and γ_n' be the corresponding cross-cuts. Since the lengths of these cross-cuts tend to zero, for every n and all sufficiently large m, γ_m' cannot be inside γ_n and vice versa. Since f is a homeomorphism, the same is true for $f(\gamma_n)$ and $f(\gamma_m')$. This proves that the chains $(f(\gamma_n))$ and $(f(\gamma_n'))$ are not equivalent.

Finally, let (c_n) be a null chain in Ω. As we have seen in the proof of Theorem 2.20, preimages $f^{-1}(c_n)$ also form a chain of cross-cuts. We would like to show that this is a null chain. Let us consider $I = \bigcap \bar{D}_n$ where $\bar{D}_n$ is the closure of the inside of $f^{-1}(c_n)$. By construction, I is the intersection of nested closed sets intersecting $\partial\mathbb{D}$; hence it is non-empty. Moreover, its intersection with the boundary of the unit disc is also a non-empty closed boundary arc. Let ω be one of the endpoints of this arc and γ_n be the family of cross-cuts converging to ω constructed as in the definition of extension of f to ω.

Let us assume that there is $z_0 \in I$ such that $|z_0 - \omega| > 0$, which is equivalent to $I \neq \{\omega\}$. This implies that the diameters of $f^{-1}(c_n)$ do not tend to zero. Since ω is on the boundary of $I \cap \partial\mathbb{D}$, $f^{-1}(c_n)$ intersects any neighbourhood of ω for all sufficiently large n. This means that if we take sufficiently large m, then $f^{-1}(c_n)$ intersects γ_m and γ_{m+1} for all sufficiently large n. Hence, c_n intersects $f(\gamma_m)$ and $f(\gamma_{m+1})$ for all sufficiently large n. Since the diameters of c_n tend to zero, this implies that the distance between $f(\gamma_m)$ and $f(\gamma_{m+1})$ is zero. This is impossible since the sequence $(f(\gamma_k))$ is a null chain by construction. This contradiction implies that $I = \{\omega\}$ and $\operatorname{diam} f^{-1}(c_n) \to 0$. In other words, $(f^{-1}(c_n))$ is a null chain with impression $\{\omega\}$. It is clearly equivalent to the null chain (γ_n). This proves that our extension is indeed the natural one and that it is onto. □

Example 2.35. With the help of prime ends, it is now easy to study properties of a conformal map which maps $\mathbb{D}$ onto the domain Ω as in Figure 2.5(a). Let f be any such map and let $\zeta = e^{i\theta}$ be the point on the boundary which is mapped to the prime end corresponding to the left side of the rectangle (denoted by I). Since all other prime ends in Ω are accessible, the function f is continuous up to the boundary everywhere except ζ. At ζ, it is not continuous since otherwise the entire boundary of Ω would be accessible.

Let u and v be the real and imaginary parts of f, respectively. Note that, as $z \to \zeta$, $f(z)$ must approach I, and hence u is continuous at ζ as well. This implies that v is discontinuous at ζ. This gives us an example of a function u which is harmonic in $\mathbb{D}$ and continuous up to the boundary, but its harmonic conjugate v is not continuous.

Remark 2.36. Finally, we would like to mention that we developed the theory of prime ends for simply connected domains. Since this theory is essentially local, with minimal modifications, it could be generalized to multiply connected domains. In particular, a theorem analogous to Theorem 2.34 is valid in the multiply connected case.

2.8. Dirichlet Boundary Problem

In this section, we discuss the connection between the Dirichlet boundary problem and conformal maps. This is a very rich subject and we are not aiming at a comprehensive cover. The main aim of this section is to show that this connection does exist and is an important one. It can be used in both directions: knowing the potential theory and the existence of the solution to the Dirichlet boundary problem, we can prove the Riemann mapping theorem and prove some results of the boundary correspondence.

Alternatively, we can prove some theorems about boundary problems using conformal map techniques.

It is important to note that the corresponding potential theory is more general in some sense. In particular, it works in every dimension, not only in the plane like complex analysis. There is a third approach which is completely ignored here: connection with stochastic analysis. A very accessible explanation of this connection between all three areas can be found in a book by Chung (2002), *Green, Brown, and Probability & Brownian Motion on the Line*. For a comprehensive cover of the connection between probability and potential theory, we refer to Doob's (2012) book, *Classical Potential Theory and its Probabilistic Counterpart: Advanced Problems.*

The cornerstone of this section is a simple observation that a composition of a harmonic function and an analytic function is harmonic. The precise statement is given by the following proposition. The proof is left to the readers.

Proposition 2.37. *Let $h : \Omega \to \mathbb{R}$ be a harmonic function in domain $\Omega \subset \mathbb{C}$ and f be an analytic function $f : \Omega' \to \Omega$. Then the function $g = h \circ f$ is harmonic in Ω'.*

We use these ideas to solve the Dirichlet boundary problem: *Given a function f on the boundary of a domain Ω, find a function u which is harmonic in Ω and has boundary values f.* This is not a particularly good formulation, since it is not clear what we mean by 'has boundary values'. For now, we interpret this as a guideline and we will discuss precise statements later on. First, we discuss a particularly simple case of the unit disc.

Throughout this section, we assume that Ω is bounded, otherwise we can use elementary functions as in the first step of the proof of the Riemann mapping theorem to map Ω onto a bounded domain Ω'. If we can find the Riemann map or solve the Dirichlet boundary problem in Ω', then we can do the same in Ω. This shows that, indeed, without loss of generality, we can assume that Ω is bounded. All definitions and constructions in this section will implicitly assume the boundedness of Ω.

2.8.1. *Poisson and Schwarz kernels*

Disc case. We start by deriving the Poisson formula which expresses the values of a harmonic function inside a disc in terms of its values on the boundary. There are many different ways to derive this formula; here, we present one of the simplest, which relies on conformal invariance (Proposition 2.37).

We start by recalling one of the most fundamental results of harmonic function: mean value property. Let u be a harmonic function in some domain Ω containing the closed disc $\{z : |z - z_0| \leq r\}$; then

$$u(z_0) = \frac{1}{2\pi} \int_0^{2\pi} u(re^{i\theta})\mathrm{d}\theta = \frac{1}{2\pi r} \int_{|\zeta - z_0| = r} u(\zeta)|\mathrm{d}\zeta|. \tag{2.3}$$

In other words, the value of harmonic function in the centre of a disc is determined by its values on the boundary of the disc. Conformal invariance implies that this is all we need to identify *all* values inside the disc.

For simplicity, let us assume that $z_0 = 0$ and $r = 1$, that is, we are working with the unit disc. Let us consider a function defined by

$$u_{z_0}(z) = u\left(\frac{z + z_0}{1 + z\bar{z}_0}\right) = u(f_{z_0}(z)).$$

This is a composition of u with a Möbius automorphism $f_{z_0} : \mathbb{D} \to \mathbb{D}$ which sends 0 to z_0. As we discussed above, this function is also harmonic in the unit disc and by mean value formula (2.3), we have

$$u(z_0) = u_{z_0}(0) = \frac{1}{2\pi} \int_{|\xi|=1} u(f_{z_0}(\xi))|\mathrm{d}\xi|.$$

Changing variables to $\zeta = f_{z_0}(\xi)$, we have

$$u(z_0) = \frac{1}{2\pi} \int_{|\zeta|=1} u(\zeta)|g'_{z_0}(\zeta)||\mathrm{d}\zeta| = \frac{1}{2\pi} \int_{|\zeta|=1} u(\zeta)\frac{1 - |z_0|^2}{|\zeta - z_0|^2}|\mathrm{d}\zeta|, \tag{2.4}$$

where g_{z_0} is the inverse of f_{z_0}. The last factor is called the *Poisson kernel* in the unit disc and denoted by $P(\zeta, z)$:

$$P(\zeta, z) = \frac{1 - |z|^2}{|\zeta - z|^2}. \tag{2.5}$$

Using this notation, formula (2.4) becomes

$$u(z_0) = \frac{1}{2\pi} \int_{|\zeta|=1} u(\zeta)P(\zeta, z_0)|\mathrm{d}\zeta|.$$

Note that the Poisson kernel is the real part of a simple function:

$$P(\zeta, z) = \frac{1 - |z|^2}{|\zeta - z|^2} = \mathrm{Re}\left(\frac{\zeta + z}{\zeta - z}\right). \tag{2.6}$$

The function $(\zeta + z)/(\zeta - z)$ is called the *Schwarz kernel.*

Sometimes, it is useful to write the Poisson kernel in polar coordinates:

$$P(\zeta, z) = P(\phi - \theta, r) = \frac{1 - r^2}{1 - 2r\cos(\phi - \theta) + r^2},$$

where $z = re^{i\phi}$ and $\zeta = e^{i\theta}$.

For a disc of radius R, the formula can be easily rescaled to

$$P(\zeta, z) = \frac{R^2 - r^2}{|\zeta - z|^2} = \frac{R^2 - r^2}{R^2 - 2rR\cos(\phi - \theta) + r^2},$$

where $z = re^{i\phi}$ and $\zeta = Re^{i\theta}$.

As mentioned above, there are many other ways of deriving the Poisson formula (2.4). For example, it can be obtained using the Fourier series expansion of the boundary values (see, for example, Gamelin (2001, Chapter X) or Rudin (1987, Section 5.4)).

In the above argument, we used the fact that u is harmonic in some neighbourhood of the unit circle. It turns out that this is not really necessary. The following proposition implies that if a function is harmonic inside the unit disc and continuous up to the boundary, then formula (2.4) still holds. It can be used to solve the Dirichlet boundary problem.

Proposition 2.38. *Let $f(\zeta)$ be a continuous function on the unit circle and define $u(z)$ inside $\mathbb{D}$ by formula* (2.4). *Then the function u is harmonic in $\mathbb{D}$ and $u(z) \to f(\zeta)$ as $z \to \zeta$ in $\mathbb{D}$.*

This integral obviously makes sense; moreover, it defines a harmonic function in $\mathbb{D}$. Indeed, the Poisson kernel is a harmonic function with respect to z since it is the real part of an analytic function. It is straightforward to check that u is harmonic, since P is harmonic. Finally, we note that the integral of the Poisson kernel with respect to ζ is equal to 2π for every z and as $z \to \zeta$, P concentrates near ζ. From this, it is not too difficult to prove that $u(z) \to f(\zeta)$ as $z \to \zeta$.

Exercise 2.13. Complete the proof of Proposition 2.38.

Since the Poisson kernel is the real part of an analytic function, the previous result can be complexified.

Proposition 2.39 (Schwarz formula). *Let f be a function analytic in the unit disc which is continuous in the closed disc. Then*

$$\begin{aligned} f(z) &= \frac{1}{2\pi}\int_{|\zeta|=1}\frac{\zeta+z}{\zeta-z}\,\mathrm{Re}\,(f(\zeta))\,|\mathrm{d}\zeta| + iC \\ &= \frac{1}{2\pi i}\int_{|\zeta|=1}\frac{\zeta+z}{\zeta-z}\,\mathrm{Re}(f(\zeta))\frac{\mathrm{d}\zeta}{\zeta} + iC \\ &= \frac{1}{2\pi}\int_0^{2\pi}\frac{e^{i\theta}+z}{e^{i\theta}-z}\,\mathrm{Re}(f(e^{i\theta}))\mathrm{d}\theta + iC, \end{aligned} \tag{2.7}$$

where $C = \mathrm{Im}(f(0))$ is a real constant.

Proof. Let us consider the function

$$u(z) = \frac{1}{2\pi}\int \mathrm{Re}\left(\frac{\zeta+z}{\zeta-z}\right)\mathrm{Re}(f(\zeta))|\mathrm{d}\zeta|.$$

By the Poisson formula, this function is harmonic inside $\mathbb{D}$ and equal to $\mathrm{Re}\, f$ on the boundary; hence, by the maximum modulus principle, it is equal to the real part of f everywhere in $\mathbb{D}$. But this function can be written as the real part of

$$\frac{1}{2\pi}\int \frac{\zeta+z}{\zeta-z}\,\mathrm{Re}(f(\zeta))|\mathrm{d}\zeta|,$$

which is also an analytic function since the Schwarz kernel is analytic as a function of z. Two analytic functions with the same real part can only differ by a purely imaginary constant. By considering $z = 0$, we get the value of C. □

In this proposition, we assumed that f is continuous up to the boundary. This condition could be relaxed. In particular, in Chapter 3, we prove the Herglotz theorem (Theorem 3.24) where we only assume that $\mathrm{Re}\, f > 0$.

This construction shows that we can solve the Dirichlet boundary problem in $\mathbb{D}$ with arbitrary continuous boundary values. Essentially, the same argument shows that if f is integrable, then $u(z) \to f(\zeta)$ as $z \to \zeta$ for every point ζ where f is continuous. Clearly, we cannot have this property at a point where f is discontinuous and we need the boundary values to be integrable, otherwise we cannot have the mean value property. Overall, this shows that the Dirichlet problem in $\mathbb{D}$ can be solved in all cases where we expect the solution to exist.

For other simply connected domains, we can use the conformal invariance of the problem and the results of boundary correspondence to solve the Dirichlet boundary problem. One possibility is to transfer the problem to the unit disc by applying the Riemann pap, solve it in $\mathbb{D}$, and transfer the answer back. Alternatively, it is possible to transfer the Poisson kernel and obtain a function $P_\Omega(\zeta, z)$ such that the solution of the Dirichlet boundary problems in Ω is given by

$$u(z) = \int_{\partial\Omega} P(\zeta, z) f(\zeta) |\mathrm{d}\zeta|.$$

In the same way as before, the Poisson kernel $P(\zeta, z)$ in a general domain Ω can be written as the real or imaginary part of an analytic function which is called the Schwarz kernel in Ω. This allows one to write an analytic function f in Ω as an integral involving the real or imaginary part of f on the boundary and the Schwarz kernel. We will not discuss the general theory, but will consider the particularly important case of $\Omega = \mathbb{H}$.

Half-plane case. Here, we briefly sketch how one can derive the half-plane version of Poisson and Schwarz formulas. We are not aiming at providing complete proof, but rather want to explain how the general strategy of obtaining these formulas for general domains works. In particular, we assume that all integrals appearing in this argument converge, without going into a detailed discussion of when it is true.

Let u be a bounded harmonic function in the upper half-plane $\mathbb{H}$ which is continuous up to the boundary. Let $z_0 = x_0 + iy_0$ be a fixed point in $\mathbb{H}$. We can define a harmonic function $\tilde{u}$ in the unit disc by

$$\tilde{u}(z) = u\left(\frac{z\bar{z}_0 - z_0}{z - 1}\right).$$

This function is harmonic as a composition of a harmonic function and a Möbius transformation $\mathbb{D} \to \mathbb{H}$ which maps 0 to z_0. Applying the mean value property (2.3) to $\tilde{u}(0)$, we obtain

$$u(z_0) = \tilde{u}(0) = \frac{1}{2\pi} \int_{|\zeta|=1} \tilde{u}(\zeta) |\mathrm{d}\zeta| = \frac{1}{\pi} \int_{-\infty}^{\infty} u(x) \frac{y_0}{(x - x_0)^2 + y_0^2} \mathrm{d}x, \quad (2.8)$$

where we have used the change of variables $\zeta = (x - z_0)/(x - \bar{z}_0)$.

This suggests that the corresponding Poisson kernel in $\mathbb{H}$ is

$$P_{\mathbb{H}}(z,t) = P(z,t) = \frac{1}{\pi}\frac{y}{|z-t|^2} = \operatorname{Im}\left(\frac{1}{\pi(t-z)}\right). \tag{2.9}$$

Note that, unlike in the disc case, the factor involving π is incorporated into the kernel.

If the function u is not bounded, then (2.8) might be wrong even if the integral converges. A typical example is $u(z) = \operatorname{Im}(z)$. This function is identically equal to 0 on the boundary, so the integral formula gives zero value for all z_0. But if we take some continuous function f on $\mathbb{R}$ (it must decay sufficiently fast, so that the integral below converges) and define u in $\mathbb{H}$ by

$$u(z) = \int_{-\infty}^{\infty} f(t)P(z,t)\mathrm{d}x,$$

then it is possible to use the same argument as in Proposition 2.38 to show that u is harmonic and its boundary values are equal to f. The problem is that if f comes from the boundary values of some other harmonic function, then we cannot conclude that two harmonic functions are equal. To do this, we also have to show that they have the same behaviour at infinity.

Finally, since P is the imaginary part of an analytic function, we can derive the half-plane version of the Schwarz formula. Because of the aforementioned problem with infinity, we formulate it for analytic functions that vanish sufficiently fast at infinity.

Proposition 2.40. *Let f be an analytic function in the upper half-plane $\mathbb{H}$, continuous up to the boundary, and such that $f(z) = O(|z|^{-\alpha})$ as $z \to \infty$ for some $\alpha > 0$. Then*

$$f(z) = \frac{1}{\pi}\int_{-\infty}^{\infty} \frac{\operatorname{Im}(f(t))}{t-z}\mathrm{d}t. \tag{2.10}$$

Proof. The proof is almost identical to the proof of Proposition 2.39. One has to show that the integral in (2.10) defines an analytic function which vanishes at infinity and whose imaginary part on the real part is the same as that of f. Since both functions vanish at infinity, this implies that they are equal. The details are left to the reader. □

Alternatively, this formula could be derived from the Cauchy integral formula. It is also possible to prove a similar formula in terms of the real part of the boundary values (similar to the disc case).

Exercise 2.14. Let f be as in Proposition 2.40. Use the Cauchy formula to show directly that (2.10) holds. Show that

$$f(z) = \frac{1}{\pi i} \int_{-\infty}^{\infty} \frac{\mathrm{Re}(f(t))}{t - z} dt.$$

Finally, the Schwarz formula in the half-plane could be obtained by composing the disc version with a conformal transformation, in the same way as we obtained the Poisson kernel.

2.8.2. *Green's function*

The Poisson kernel is extremely important when we want to solve the boundary problem for $\Delta u = 0$. When we are interested in solutions to $\Delta u = f$, we need Green's function, which we define by its properties.

Definition 2.41. We say that $G_\Omega(z_0, z) = G(z_0, z)$ is Green's function in a domain Ω with the pole at z_0 if

- G is harmonic with respect to z in $\Omega \setminus \{z_0\}$,
- $G(z_0, z) = g(z_0, z) - \log|z - z_0|$ where g is a function which is harmonic in the entire Ω,[7]
- $G(z_0, z) \to 0$ as $z \to \partial\Omega$.

It is not obvious that such a function exists, but from the maximum modulus principle, it is immediately clear that Green's function must be unique. Since G is equal to zero on the boundary and blows up to infinity near z_0, by the maximum modulus principle, it must be positive in G.

The existence of Green's function is closely related to the solution of the Dirichlet boundary problem. Indeed, let $g(z_0, z)$ be a harmonic function (with respect to z) in Ω with boundary values given by $\log|z - z_0|$ for $z \in \partial\Omega$. Then $G(z_0, z) = g(z_0, z) - \log|z - z_0|$ is Green's function with the pole at z_0.

Example 2.42. The simplest case is $\Omega = \mathbb{D}$ and $z_0 = 0$. In this case, it is easy to check that Green's function with the pole at $z_0 = 0$ is

$$G_{\mathbb{D}}(z_0, z) = -\log|z| = \log\frac{1}{|z|}.$$

[7]In the context of the fundamental solution to $\Delta u = f$, the usual definition of Green's function differs from ours by the factor of $1/2\pi$. This does not change much, but one has to be careful with the coefficients in various formulas.

For poles at other points, we can use the Poisson formula (2.4) to solve the corresponding boundary problem, or, alternatively, we could apply conformal invariance directly to G.

Since harmonic functions are conformally invariant, the same is true for Green's function, namely, if $f : \Omega \to \Omega'$, then

$$G_{\Omega'}(f(z_0), f(z)) = G_\Omega(z_0, z). \tag{2.11}$$

The proof is straightforward; we only have to take care of the logarithmic singularity at z_0. The details are left as an exercise.

Exercise 2.15. Complete the proof of conformal invariance of Green's function.

Let $f : \Omega \to \mathbb{D}$ be a Riemann map with $f(z_0) = 0$. Combining (2.11) with the explicit formula for Green's function in $\mathbb{D}$, we get

$$G_\Omega(z_0, z) = \log \frac{1}{|f(z)|}.$$

This proves that Green's function can be found using the Riemann map.

Example 2.43. Applying (2.11) to $\Omega = \Omega' = \mathbb{D}$ and $f(z) = (z - z_0)/(1 - z\overline{z}_0)$, we get

$$G_\mathbb{D}(z_0, z) = G_\mathbb{D}(0, f(z)) = -\log|f(z)| = \log\left|\frac{1 - z\overline{z}_0}{z - z_0}\right|.$$

Finally, we want to show that the Riemann map can be recovered from Green's function. We know that $G = -\log|f| = -\operatorname{Re}\log(f)$, so it would be natural to say that $f = \exp(-G - i\tilde{G})$, where $\tilde{G}$ is a harmonic conjugate of G. The problem is that G is harmonic only in $\Omega \setminus \{z_0\}$, which is not simply connected, and we do not know whether there is a single-valued harmonic conjugate or not.

Instead, we write $G(z_0, z) = -\log|z - z_0| + g(z_0, z)$ and note that $g(z) = g(z_0, z)$ is harmonic in a simply connected domain; hence it has a harmonic conjugate which is single-valued and unique up to an additive real constant. Let us choose this constant in such a way that $\tilde{g}(z_0) = 0$. Consider a function

$$f(z) = (z - z_0)e^{-g(z) - i\tilde{g}(z)}.$$

This function is obviously analytic; $f(z_0) = 0$ and $f'(z_0) = e^{-g(z_0)} > 0$. Moreover, on the boundary of Ω, it is of modulus 1. Note that $|f| \to 1$ as $z \to \partial\Omega$, but we do not know that f is continuous at the boundary since we have no control over continuity of $\tilde{g}$. In fact, it might happen that it is discontinuous (see Example 2.35).

It remains to be proven that f is one-to-one. Together with the boundary values of $|f|$, we see that $f(\Omega) = \mathbb{D}$. The proof of injectivity of f is very similar to the proof of Theorem 2.28, which, unfortunately, cannot be directly applied here since $\partial\Omega$ is not necessarily a Jordan curve.

We claim that there are simply connected domains Ω_n such that $\Omega_n \subset \Omega_{n+1}$, Ω_n converges to Ω (that is, $\Omega = \bigcup \Omega_n$), the Ω_n are bounded by simple closed Jordan curves γ_n, and γ_n is in 2^{-n}-neighbourhood of $\partial\Omega$. There are many ways to show that this is true. Probably the simplest is to consider all the dyadic squares with side length 2^{-n} that are inside Ω and take the interior of the component of their union which contains z_0.

The modulus of f converges uniformly to 1 on γ_n, that is, $r_n = \inf_{\gamma_n} |f(z)| \to 1$. If this is not so, then for some ε, there are points $z_n \in \gamma_n$ such that $|f(z_n)| < 1 - \varepsilon$ and we can choose a subsequence such that $|f(z_{n_k})| \to c \leq 1 - \varepsilon$. By construction, all points of accumulation of the sequence $\{z_{n_k}\}$ are on the boundary of Ω and we get a contradiction since we know that $|f| \to 1$ as $z \to \partial\Omega$.

Let us consider a function

$$\mathrm{ind}_n(w) = \frac{1}{2\pi i} \int_{\gamma_n} \frac{f'(z)}{f(z) - w} \mathrm{d}z,$$

which gives the index of $f(\gamma_n)$ with respect to w, that is, how many times the image of γ_n goes around w in the counterclockwise direction. This function is well defined and analytic for all $|w| < r_n$. It is also integer-valued, so $\mathrm{ind}_n(w) = \mathrm{ind}_n(0)$ for all $|w| < r_n$. By the argument principle, $\mathrm{ind}_n(w)$ also gives the number of solutions of $f(z) = w$ for $z \in \Omega_n$. By construction, we have that $f(z) = 0$ if and only if $z = z_0$. This proves that $\mathrm{ind}_n(w) = 1$ for all $|w| < r_n$. Passing to the limit as $n \to \infty$, we have that f is one-to-one and onto.

Remark 2.44. This gives an alternative way of proving the Riemann mapping theorem which is, in fact, very close to Riemann's original proof. It is possible to show that the Dirichlet boundary problem has a solution without the use of conformal maps. The standard method is due to Perron and can be found in many books on complex analysis or potential theory. We refer interested readers to Ahlfors (1978, Chapter 6, Section 4).

2.8.3. *Harmonic measure*

The harmonic measure is one of the fundamental objects in geometric function theory and plays an important role in many applications. In particular, it can be used to prove the multiply connected version of the Riemann mapping theorem (see Ahlfors (1978, Section 6.5)). Extensive discussion of the harmonic measure can be found in Garnett and Marshall (2008).

There are several ways to define the harmonic measure. Here, we will present some of them, but we will not prove that they all are equivalent. Probably, the simplest way to form a definition is via conformal invariance.

Definition 2.45. For the unit disc, we define the harmonic measure $\omega_{\mathbb{D}}(0, A)$ on the boundary of $\mathbb{D}$ as the normalized Lebesgue measure $m(A)/2\pi$. For any simply connected domain Ω and $z_0 \in \Omega$, we define $\omega_\Omega(z_0, A)$ as the harmonic measure of A with respect to z_0 (or with the pole at z_0) as $\omega_{\mathbb{D}}(0, f(A))$, where f is a conformal map from Ω onto $\mathbb{D}$ with $f(z_0) = 0$. We understand $f(A)$ in terms of prime ends. When A is not a subset of the boundary, we interpret $\omega_\Omega(z_0, A)$ as $\omega_\Omega(z_0, A \cap \partial\Omega)$.

Conformal invariance is built into this definition. An alternative definition uses the Dirichlet boundary problem. This definition makes sense for multiply connected domains and in higher dimension.

Definition 2.46. Let Ω be a simply connected domain and A be a set on its boundary. The harmonic measure $\omega_\Omega(z, A)$ is defined as $u(z)$, where u is the solution of the Dirichlet boundary problem with the boundary value $u = 1$ on A and $u = 0$ on the rest of the boundary.

It is not difficult to check that these two definitions are equivalent. The main difference is that in the first definition, we mainly consider $\omega(z, A)$ as a measure which depends on a parameter z. In the second definition, we assume that it is a harmonic function of z which depends on a parameter A.

Another definition in terms of the Dirichlet boundary problem is to set $\omega_\Omega(z_0, \cdot)$ as the measure which solves the boundary problem: for every function u which is harmonic in Ω and continuous up to the boundary, we have

$$u(z_0) = \int_{\partial\Omega} u(\zeta)\omega_\Omega(z_0, \mathrm{d}\zeta).$$

This suggests that one might assume that the harmonic measure is a generalization of the Poisson kernels. Indeed, it is true that if the boundary of Ω is smooth, then the harmonic measure is absolutely continuous with respect to the arc length and its density is exactly the Poisson kernel (see Exercise 2.16).

Readers familiar with Brownian motion might find the following definition more illustrative.

Definition 2.47. Let Ω be a domain, A be a set on its boundary and B_t be a standard two-dimensional Brownian motion started from z. The harmonic measure of A with respect to z can be defined as $\omega_\Omega(z, A) = \mathbb{P}(B_\tau \in A)$, where $\tau = \inf\{t > 0 : B_t \notin \Omega\}$ is the first exit time.

One of the main simple properties of harmonic measure is that it is monotone with respect to both Ω and A. The precise statement is given by the following theorem.

Theorem 2.48. *Let Ω be a subdomain of Ω'. Let us assume that $A \subset (\partial\Omega \cap \partial\Omega')$ and that $z \in \Omega$. Then $\omega_\Omega(z, A) \leq \omega_{\Omega'}(z, A)$. If $A \subset A' \subset \partial\Omega$, then $\omega_\Omega(z, A) \leq \omega_\Omega(z, A')$.*

Proof. Both parts of the theorem follow from the maximum principle for harmonic functions. Obviously, $h(z) = \omega_{\Omega'}(z, A)$ is a harmonic function in Ω; moreover, it dominates $\omega_\Omega(z, A)$ on the boundary of Ω. Indeed, the boundary of Ω is made up of three parts: A, $(\partial\Omega \cap \partial\Omega') \setminus A$, and $\partial\Omega \cap \Omega'$. On the first two, both harmonic measures are equal to 1 and 0, respectively. On the last part, the harmonic measure in Ω is equal to 0 and the harmonic measure in Ω' is non-negative.

The second inequality is proved in a similar way. Indeed, considering the boundary values, we see that $\omega_\Omega(z, A) + \omega_\Omega(z, A' \setminus A) = \omega_\Omega(z, A')$. As before, $\omega_\Omega(z, A' \setminus A) \geq 0$ and the desired inequality follows immediately.

We can also note that both inequalities are strict unless $\Omega = \Omega'$ or the harmonic measure of $A' \setminus A$ is identically equal to 0. □

Exercise 2.16. The harmonic measure in the upper half-plane is continuous with respect to the Lebesgue measure and its density is given by the Poisson kernel, namely, the density of $\omega_{\mathbb{H}}(z, t)$ is

$$\frac{1}{\pi} \frac{y}{y^2 + (x - t)^2},$$

where $z = x + iy$.

2.9. Multiply Connected Domains

2.9.1. *Conformal annuli*

In the previous section, we have shown that all non-trivial simply connected domains are conformally equivalent to the unit disc, and hence they all are conformally equivalent to each other. For multiply connected domains, this is no longer true. The simplest example is given by the following theorem.

Theorem 2.49. *Let $A(r, R) = \{z : r < |z| < R\}$ be an annulus with the smaller radius r and the larger radius R. In the case $0 < r_i < R_i < \infty$, there is a conformal map from $A_1 = A(r_1, R_2)$ onto $A_2 = A(r_2, R_2)$ if and only if $R_1/r_1 = R_2/r_2$. For degenerated annuli (where $r_i = 0$ and/or $R_i = \infty$), the situation is more complicated. The annulus $A(0, \infty)$ is not conformally equivalent to any other annulus and all annuli $A(0, R)$ and $A(r, \infty)$ with $r > 0$ and $R < \infty$ are conformally equivalent to each other and not equivalent to any other type of annuli.*

Proof. If $0 < r_i < R_i < \infty$ and ratios of the radii are the same, then $f(z) = zR_2/R_1$ maps $A(r_1, R_1)$ onto $A(r_2, R_2)$. This map is linear and hence conformal. The same function maps $A(0, R_1)$ onto $A(0, R_2)$, zr_2/r_1 maps $A(r_1, \infty)$ onto $A(r_2, \infty)$ and, finally, $R_1 r_2/z$ maps $A(0, R_1)$ onto $A(r_2, \infty)$. This shows that all annuli that are supposed to be conformally equivalent are indeed conformally equivalent.

First, we consider the non-degenerate case where, without loss of generality, we can assume that $r_i = 1$.

The main part of the theorem is the statement that the ratio of radii is a conformal invariant: it is not changed by a conformal map. Let us assume that there is a map f from one annulus onto another. We will show that this implies that the ratios of radii are equal.

First, we want to show that f maps boundary circles onto boundary circles. Note that this is much weaker than continuity up to the boundary, and this is why we can show this without the use of sophisticated techniques.

Let $S = r\mathbb{T}$ be a circle in A_2 with radius $r \in (1, R_2)$. Its preimage under f is a compact set; hence it is bounded away from both boundary circles of A_1. In particular, $K = f(A(1, 1+\varepsilon))$ does not intersect S for sufficiently small ε. Since S separates A_2 into two disjoint parts, this means that K is completely inside S or completely outside S (Figure 2.14). Let us assume for a while that it is inside. If we consider a sequence $\{z_n\}$ inside $A(1, 1+\varepsilon)$ with $|z_n| \to 1$, then the sequence $\{f(z_n)\}$ does not have points of accumulation

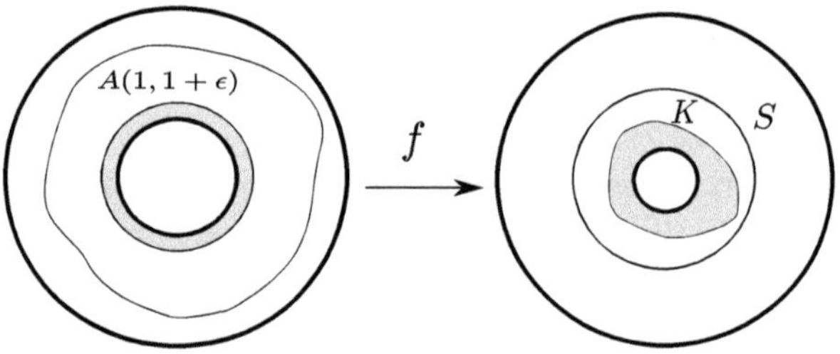

Fig. 2.14. Circle S and its preimage split each annulus into two doubly connected domains. The shaded areas are $A(1, 1+\varepsilon)$ and its image. We assume that both of them lie inside S and its preimage.

inside A_2, and hence $|f(z_n)|$ must converge to 1. In the same way, we show that $|f(z_n)| \to R_2$ for $|z_n| \to R_1$. The purpose of the trick with S is to exclude the possibility that $f(z_n)$ oscillates between two boundary circles.

In the case when K is outside S, we get that $|f(z_n)| \to 1$ as $|z_n| \to R_1$ and $|f(z_n)| \to R_2$ as $|z_n| \to 1$. In this case, we change $f(z)$ to $R_2/f(z)$, which also conformally maps A_1 onto A_2 but has the same boundary behaviour as in the first case.

Let us consider the function

$$u(z) = \log|f(z)| = \mathrm{Re}\,(\log(f(z)))\,.$$

Since $|f|$ is real and positive, u is a well-defined, single-valued function. It is also a real part of an analytic function and hence it is harmonic in A_1. The previous discussion shows that u can be extended continuously to the closure of A_1 by defining $u(z) = 0$ on $|z| = 1$ and $u(z) = \log(R_2)$ on $|z| = R_1$. There is another harmonic function in A_1 which has the same boundary values:

$$\frac{\log(R_2)}{\log(R_1)} \log|z|.$$

The difference of these two functions is harmonic and equal to 0 on the boundary; by the maximum modulus principle, the difference is 0 everywhere and two harmonic functions are equal.

The basic idea of the rest of the proof is very simple. The equality of the harmonic functions gives $|f| = |z|^\alpha$, where $\alpha = \log(R_2)/\log(R_1)$. This suggests that $f = cz^\alpha$ for some c with $|c| = 1$. But z^α is one-to-one if and only if $\alpha = 1$ or, equivalently, $R_1 = R_2$. The rigorous justification of this argument is slightly more involved.

Let us consider a harmonic function

$$h(z) = \log|f| - \alpha \log|z|.$$

This function looks like the real part of $\log(f) - \alpha\log(z)$, but we do not know whether it can be defined as a single-valued function.

The above argument shows that h vanishes on the boundary; hence, by the maximum principle, it vanishes everywhere in A_1. Equivalently, $\log|f|^2 = \alpha\log|z|^2$ or $\log(f\bar{f}) = \alpha\log(z\bar{z})$. Applying the Cauchy–Riemann differential operator $\partial = (\partial_x - i\partial_y)/2$ to both functions, we get

$$\frac{f'}{f} = \alpha\frac{1}{z}.$$

Take any simple curve γ which goes counterclockwise around the origin inside A_1 and integrate this identity along γ. Dividing by $2\pi i$, we have

$$\frac{1}{2\pi i}\oint_\gamma \frac{f'(z)}{f(z)}\mathrm{d}z = \alpha.$$

By the argument principle, the left-hand side is the index of $f(\gamma)$ and hence α must be an integer. This proves that z^α is a single-valued analytic function and $|f|$ is equal to its modulus. As argued above, this implies that $\alpha = 1$ and that f is a rotation.

Next, assume that $r_1 > 0$ and let $f : A(r_1, R_1) \to A(r_2, R_2)$. The same argument as above shows that $|f(z)| \to r_2$ or $|f(z)| \to R_2$ as $z \to r_1$. In the first case, we consider the Laurent expansion of f around the origin. Its coefficients are

$$a_n = \frac{1}{2\pi i}\int_{|z|=r} \frac{f(\zeta)}{\zeta^{n+1}}\mathrm{d}\zeta,$$

where r is any number between r_1 and R_1. If $r_2 = 0$, then passing to the limit as $r \to r_1$, we get that all coefficients are equal to zero, and f is not conformal. This proves that $r_2 > 0$. In the second case, we consider $1/f$ and show that $1/R_2 > 0$. In either case, one of the radii must be non-degenerate. □

This is a very important theorem and, as such, it has more than one proof. Here, we give one more proof and we will give another one after the discussion of extremal lengths in Section 4.2.1. The second proof is based on the following proposition which we state and prove only in the non-degenerate case.

Proposition 2.50. *Let A_1 and A_2 be two annuli as before. If there is a univalent map $f : A_1 \to A_2$, then $R_2/r_2 \geq R_1/r_1$.*

Proof. As before, we can assume without loss of generality that $r_1 = r_2 = 1$ and that the inner circle is mapped to the inner circle, so that the outer circle is mapped to the outer circle. Since function f is analytic in an annulus, it can be written as the Laurent series

$$f(z) = \sum_{n\in\mathbb{Z}} a_n z^n.$$

We denote by $A(r)$ the area of a domain bounded by a Jordan curve $f(re^{i\theta})$ where θ goes from 0 to 2π. By Green's formula for the area, we have

$$\begin{aligned} A(r) &= \frac{1}{2i}\int \bar{f}(z)\mathrm{d}f(z) = \frac{1}{2i}\int_{|z|=r} \bar{f}(z)f'(z)\mathrm{d}z \\ &= \frac{1}{2i}\int_0^{2\pi} \left(\sum \bar{a}_n r^n e^{-i\theta n}\right)\left(\sum n a_n r^{n-1} e^{i\theta(n-1)}\right) rie^{i\theta}\mathrm{d}\theta \\ &= \pi\sum_{n\in\mathbb{Z}} n|a_n|^2 r^{2n}. \end{aligned}$$

The last identity holds since $\int e^{in\theta} = 0$ unless $n = 0$. Passing to the limit as $r \to 1$, we have

$$\pi = \pi\sum n|a_n|^2.$$

Using this identity, we can write

$$A(r) - \pi r^2 = \pi r^2 \sum_{n\in\mathbb{Z}} n|a_n|^2(r^{2n-2} - 1) \geq 0,$$

where the last inequality holds termwise (indeed, n and $r^{2n-2} - 1$ have the same sign). Passing to the limit as $r \to R_1$, we obtain that $R_2 \geq R_1$. □

To complete the second proof of Theorem 2.49, we just use the Proposition for f and f^{-1}.

Exercise 2.17. Use reflection principle to prove Theorem 2.49 in a slightly weaker form: Let A_1 and A_2 be two annuli as before. Let us also assume that there is a conformal map $f : A_1 \to A_2$ which is continuous up to the boundary. Then A_1 and A_2 have the same shape, that is, $R_1/r_1 = R_2/r_2$. (As mentioned before, the boundary correspondence results could be used to show that the assumption of continuity up to the boundary always holds even in a doubly connected case, so our assumption is not really restrictive.)

Theorem 2.49 shows that not all doubly connected domains are conformally equivalent, and it is easy to believe that the same is true for domains of higher connectivity. This means that we cannot use the same uniformizing domain for all domains; instead, we should use sufficiently large families of standard domains. In the doubly connected case, the standard choice is the family of all annuli with outer radius 1 (some authors prefer annuli with inner radius 1). For higher connectivity, there is no standard family, but there are several preferred families. One such family is the family of circle domains: domains such that each boundary component is either a circle or a single point. For infinitely connected domains, the Koebe conjecture states that every domain can be mapped onto a circle domain. The best result in this direction is given by He and Schramm (1993) who proved this conjecture for countably connected domains using circular packing techniques. We discuss other standard families of domains at the end of this section.

We start with a general construction that works for all finitely connected domains. This allows us to assume without loss of generality that all boundary components are analytic Jordan curves.

First, we can get rid of all single point components. Indeed, if there is a map f from $\Omega \setminus \{z_0\}$, then z_0 is an isolated singularity and the function is univalent in its neighbourhood. Considering f as a $\widehat{\mathbb{C}}$-valued function, this implies that z_0 is a removable singularity. In other words, the image of a single-point boundary component must also be a single-point boundary component and both of them could be removed.

To show that we can assume all boundary components are 'nice', we again use the Riemann uniformization theorem. Let Ω be an n-connected domain and let $E_1, \ldots, E_n$ be the components of its complement. Using the above argument, we assume that all E_i are not singletons. Let us consider the domain $\Omega \cup E_2 \cup \cdots \cup E_n$. This is a simply connected domain whose complement is not a single point; hence we can map it to the unit disc. Under this map $\Omega, E_2, \ldots, E_n$ are mapped to some subsets of $\mathbb{D}$ which, abusing notation, we still call $\Omega, E_2, \ldots, E_n$. We denote the complement of the unit disc by new E_1. Note that the boundary of E_1 is now the unit circle, which is an analytic Jordan curve. Next, we take the union of all domains except E_2, map it to the disc and rename all the sets. After this, the boundary of E_2 is the unit circle and the boundary of E_1 is a univalent image of the unit circle, and hence it is an analytic Jordan curve. Continuing in this way for all components, we can map the original domain onto a subdomain of $\mathbb{D}$ such that one boundary component is the unit circle and the others are analytic Jordan curves (see Figure 2.15).

If all boundary components of the original domain were locally connected then, by the Carathéodory theorem, the map that we have constructed is continuous up to the boundary in the sense of accessible points.

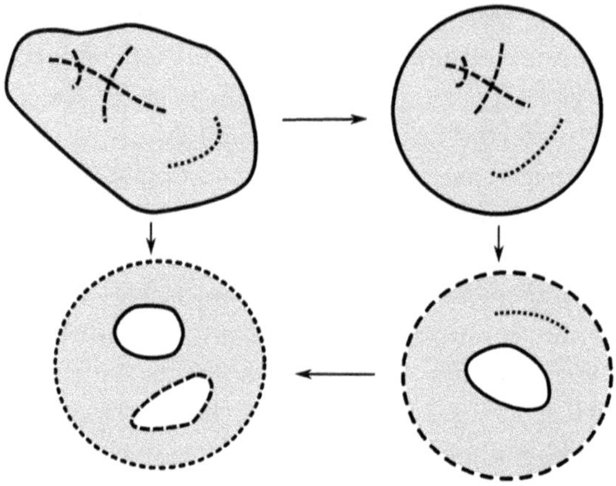

Fig. 2.15. Dashed, dotted, and solid lines represent three boundary components and their successive images.

If all boundary components were Jordan curves, then the map is continuous up to the boundary in the usual sense.

We start the treatment of multiply connected uniformization by presenting a rather elementary proof for doubly connected domains. The general case will be treated later using a completely different approach.

Theorem 2.51. *Let Ω be a doubly connected domain such that at least one of the components of its complement contains more than one point. Then there is a univalent map f from Ω onto some annulus with outer radius* 1. *This map is unique up to rotation and inversion of the annulus.*

If both components contain only one point, then Ω is conformally equivalent to the punctured plane $\mathbb{C} \setminus \{0\} = \widehat{\mathbb{C}} \setminus \{0, \infty\} = A(0, \infty)$.

Proof. We have already proved uniqueness in the proof of Theorem 2.49. To prove existence, we first consider two special cases. If both components of the complement are singletons, then we can choose f to be a Möbius transformation sending these two points to 0 and ∞. If only one of them is a single point, then we can map Ω with this point to the unit disc and this point to the origin.

The only interesting case is when both components are non-trivial. As explained above, we can assume that Ω is a doubly connected domain such that one component of its complement is the complement of the unit disc and the other one is bounded by an analytic curve. By composing with one

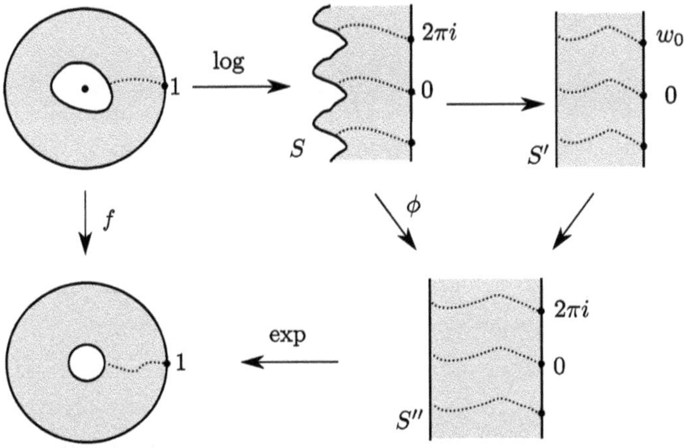

Fig. 2.16. The Riemann mapping from a doubly connected domain onto an annulus. Dashed lines form an arbitrary simple curve connecting 1 to the inner boundary component and its images.

more Möbius transformation, we can assume that the origin is inside the second component.

Let us apply the logarithmic function to Ω. Since 0 is not in Ω, the logarithm is analytic, but it is not single-valued. Each time we go around the inner boundary component, the value of log changes by $2\pi i$. The logarithmic function maps Ω onto a vertical strip S such that its right boundary is the imaginary axis and the left boundary is a $2\pi i$-periodic curve (see Figure 2.16). By the Riemann theorem, there is a univalent map from S onto a vertical strip $S' = \{z : -1 < \mathrm{Re}(z) < 0\}$. Moreover, we can assume that $\pm i\infty$ and 0 are mapped to themselves. The point $2\pi i$ is mapped to some point w_0 on the positive imaginary axis. Rescaling by $2\pi/|w_0|$, we find a map ϕ from S onto $S'' = \{z : -h < \mathrm{Re}(z) < 0\}$ where $h = 2\pi/|w_0|$. This map preserves $\pm i\infty$, 0 and $2\pi i$. We claim that ϕ satisfies the following equation:

$$\phi(z + 2k\pi i) = \phi(z) + 2k\pi i. \tag{2.12}$$

Moreover, the same is true for the inverse function. Obviously, it is sufficient to prove this for $k = 1$, as the general case follows immediately by induction. Note that $z \mapsto z + 2\pi i$ is a conformal automorphism of both S and S''; hence both $f(z)+2\pi i$ and $f(z+2\pi i)$ map S onto S'' in such a way that three boundary points $\pm i\infty$ and 0 are mapped to $\pm i\infty$ and $2\pi i$. By uniqueness of the Riemann map which sends three given boundary points to three given boundary points, these two maps are the same. The proof for the inverse function is exactly the same.

Finally, we consider

$$f(z) = e^{\phi(\log(z))}.$$

This is an analytic function which maps Ω onto an annulus $A(e^{-h}, 1)$. The problem is that both log and exp are not one-to-one, so we cannot immediately claim that f is univalent. Despite that, this function is univalent. This function is injective since log maps z onto a $2\pi i$-periodic sequence. By (2.12), ϕ maps $2\pi i$-periodic sequences to $2\pi i$-periodic sequences, and, finally, exp maps any $2\pi i$-periodic sequence to a single point. A similar argument for inverse functions gives that f is surjective. □

Theorem 2.14 shows that all non-trivial simply connected domains are conformally equivalent to each other. Theorems 2.49 and 2.51 show that for doubly connected domains, there is a one-parameter family of equivalence classes. Each doubly connected domain is conformally equivalent to an annulus and the ratio of its radii completely determines the equivalence class. This is an important example of a *conformal invariant*: a quantity that does not change under conformal transformation. For various reasons that we will discuss later, the standard conformal invariant of a doubly connected domain Ω which describes the equivalence class is the *conformal modulus* $M(\Omega)$ which is defined as

$$\frac{1}{2\pi} \log \frac{R}{r},$$

where R and r are the smaller and larger radii of an annulus which is conformally equivalent to Ω. By Theorems 2.49 and 2.51, this quantity is well defined and does not depend on a particular choice of an annulus.

Exercise 2.18. Find an explicit map from a domain bounded by two non-concentric circles onto an annulus.

2.9.2. *Uniformization of multiply connected domains*

Annuli are the natural 'standard' doubly connected domains. For the domains of higher connectivity, there is no natural unique choice of uniformizing domains. Instead, there are several somewhat standard families of *canonical domains*. In this section, we discuss canonical domains and the corresponding uniformizing maps.

Multiply connected domains are not the main subject of this book, so we are not going to study them extensively. In particular, we are not going to discuss connections between different uniformizations or multiply connected version of the Christoffel–Schwarz formula. Here, we only briefly discuss some standard uniformization results and show how to prove one of them. We will provide most of the details, but will gloss over some technicalities about applying the Hurwitz theorem (Theorem 2.12) to functions in multiply connected domains. We provide an abridged version of the exposition from Nehari (1975, Section VII.2). We direct interested readers to Nehari's book for a detailed discussion of various uniformizing domains and functions as well as connections between them.

In this section, we always assume that Ω is a finitely connected domain in $\widehat{\mathbb{C}}$ such that its boundary components are analytic curves. This assumption is not restrictive since any domain could be mapped conformally onto a domain with analytic boundary. See Figure 2.15 and the related discussion.

Parallel Slit Domains. These are domains that are the complex sphere $\widehat{\mathbb{C}}$ without a finite union of intervals that are parallel to each other (Figure 2.17).

The main result is that for every multiply connected domain Ω and θ, there is a univalent function mapping Ω onto a parallel slit domain where all slits form an angle θ with the real axis. The precise statement is given by the following theorem.

Theorem 2.52. *Let Ω be a multiply connected domain and $z_0 \in \Omega$. If $z_0 \neq \infty$, then there is a unique function $f_{z_0,\theta} = f_\theta$ which maps Ω onto a slit domain with the slope θ such that its Laurent series at z_0 is of the form*

$$f_{z_0,\theta}(z) = \frac{1}{z - z_0} + a_\theta(z - z_0) + a_2(z - z_0)^2 + \cdots . \tag{2.13}$$

If $z_0 = \infty$, then the Laurent series is of the form

$$f_\theta(z) = z + a_\theta z^{-1} + a_2 z^{-2} + \cdots . \tag{2.14}$$

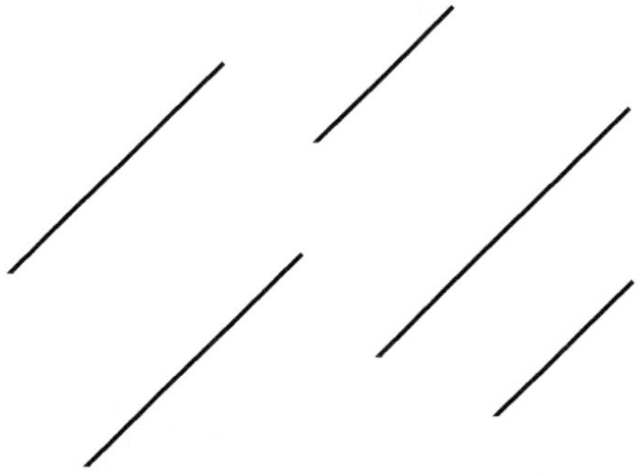

Fig. 2.17. A parallel slit domain.

There are two main ways to prove this theorem. One is using the Dirichlet boundary problem and harmonic functions, in the same spirit as the argument connecting the Riemann map and Green's function. A detailed description of this approach can be found in Ahlfors (1978, Section 6.5) or Nehari (1975, Section VII.3). Here, we present a different approach which is based on solving some optimization problem. This proof is very similar in spirit to the proof of the simply connected version of the Riemann mapping theorem from Section 2.4.

To prove this theorem, we consider $S_\theta = S_{z_0,\theta}$ the class of all functions univalent in Ω that have expansion as in (2.13) or (2.14). We state that these functions are admissible.

The main idea of the proof is to show that there is an admissible function maximizing $\mathrm{Re}(e^{-2i\theta}a_1)$. We show that this function is onto a slit domain. As we can see, this method is very similar to the proof of the Riemann mapping theorem (Theorem 2.14). The difference is in the class of admissible functions and the functional which we are trying to maximize.

The proof will use a technical result which is a direct corollary of Green's formula.

Lemma 2.53. *Let Ω be a multiply connected domain bounded by analytic curves, and let f and g be functions analytic in Ω with the possible exception of a finite number of poles in Ω. Assume that they are regular up to the boundary and that $f - g$ is analytic everywhere in Ω. Then*

$$\int_\Omega |f' - g'|^2 \mathrm{d}x\mathrm{d}y = \mathrm{Re}\left(\frac{1}{i}\int_{\partial\Omega} (\bar{f} - \bar{g}) f' \mathrm{d}z\right) + \frac{1}{2i}\int_{\partial\Omega} \bar{g} g' \mathrm{d}z + \overline{\frac{1}{2i}\int_{\partial\Omega} \bar{f} f' \mathrm{d}z}. \tag{2.15}$$

Proof. Using the complex form of Green's formula (1.1) and the fact that $\partial_{\bar{z}} \bar{f} = \bar{f}'$, we can write

$$\begin{aligned}\int_\Omega |f' - g'|^2 &= \int_\Omega (f' - g')\overline{(f' - g')} = \frac{1}{2i}\int_{\partial\Omega} (f' - g')\overline{(f - g)} \mathrm{d}z \\ &= \frac{1}{2i}\int_{\partial\Omega} \overline{(f - g)} f' \mathrm{d}z + \frac{1}{2i}\int_{\partial\Omega} \bar{g} g' \mathrm{d}z - \frac{1}{2i}\int_{\partial\Omega} \bar{f} g' \mathrm{d}z.\end{aligned}$$

We integrate the last integral by parts (the integrated term vanishes since $\partial\Omega$ is a union of closed curves)

$$\int_{\partial\Omega} \bar{f} g' \mathrm{d}z = \int_{\partial\Omega} \bar{f} \mathrm{d}g = -\int_{\partial\Omega} g \overline{\mathrm{d}f} = -\overline{\int_{\partial\Omega} \bar{g} f' \mathrm{d}z}.$$

Finally, we add and subtract the conjugate of the first term to obtain

$$\operatorname{Re}\left(\frac{1}{i}\int_{\partial\Omega}(\bar{f}-\bar{g})f'\mathrm{d}z\right) + \frac{1}{2i}\int_{\partial\Omega}\bar{g}g'\mathrm{d}z$$
$$+\frac{1}{2i}\overline{\int_{\partial\Omega}\bar{g}f'\mathrm{d}z} - \frac{1}{2i}\overline{\int_{\partial\Omega}\overline{(f-g)}f'\mathrm{d}z}.$$

The last two terms simplify to

$$\frac{1}{2i}\overline{\int_{\partial\Omega}\bar{f}f'\mathrm{d}z},$$

and we obtain (2.15). □

Let us assume for a while that a slit domain mapping f_θ does exist. Let f be any other admissible function. We can apply Lemma 2.53 to f_θ and f to obtain

$$\int_\Omega |f_\theta' - f'|^2 = \operatorname{Re}\left(\frac{1}{i}\int_{\partial\Omega}(\bar{f}_\theta - \bar{f})f_\theta'\right)$$
$$+\frac{1}{2i}\int_{\partial\Omega}\bar{f}f' - \frac{1}{2i}\overline{\int_{\partial\Omega}\bar{f}_\theta f_\theta'}. \tag{2.16}$$

Since f_θ maps onto a slit domain, $e^{-i\theta} f_\theta' \mathrm{d}z$ is real everywhere on the boundary. This allows one to write

$$f_\theta'(z)\mathrm{d}z = e^{2i\theta}\overline{f_\theta'(z)\mathrm{d}z}, \quad z \in \partial\Omega. \tag{2.17}$$

This identity can be used to simplify some integrals in the above formula. We start with the first one. Using (2.17) and the residue calculus,

$$\frac{1}{i}\int_{\partial\Omega}(\bar{f}_\theta - \bar{f})f_\theta'\mathrm{d}z = \frac{e^{2i\theta}}{i}\overline{\int_{\partial\Omega}(f_\theta - f)f_\theta'\mathrm{d}z} = \overline{(2\pi e^{-2i\theta}(a_\theta - a_1))},$$

where a_θ and a_1 are the coefficients of f_θ and f in front of $(z - z_0)$ for finite z_0 and in front of z^{-1} if $z_0 = \infty$. Thus, the first term in (2.16) simplifies to

$$2\pi \operatorname{Re}(e^{-2i\theta}(a_\theta - a_1)).$$

Using Green's formula once again, we obtain that the second term in (2.16) is $-A$ where $A = \text{Area}(\mathbb{C} \setminus \text{f}(\Omega))$ is the area of the domains inside the images of boundary components of $f(\Omega)$. The minus sign comes from the fact that the boundary is positively oriented with respect to $f(\Omega)$ instead of its complement.

Using (2.17) one more time, we simplify the last term in (2.16)

$$\int_{\partial\Omega} \bar{f}_\theta f'_\theta \mathrm{d}z = e^{2i\theta} \overline{\int_{\partial\Omega} f_\theta f'_\theta \mathrm{d}z} = \frac{e^{2i\theta}}{2} \overline{\int_{\partial\Omega} \mathrm{d}(f_\theta^2)} = 0.$$

Combining all these simplifications together, we arrive at

$$\int_\Omega |f'_\theta - f'|^2 = 2\pi \,\mathrm{Re}(e^{-2i\theta}(a_\theta - a_1)) - A.$$

Since $A \geq 0$, we have that $\mathrm{Re}(e^{-2i\theta} a_\theta) \geq \mathrm{Re}(e^{-2i\theta} a_1)$. Moreover, the equality happens if and only if

$$\int_\Omega |f'_\theta - f'|^2 = 0,$$

which happens if and only if f and f_θ differ by a constant. In particular, this implies that if a normalized map onto a slit domain exists, then it must be unique.

Applying this argument to $f(z) = 1/(z - z_0)$ for finite z_0 or $f(z) = z$ if $z_0 = \infty$, we can see that $\mathrm{Re}(e^{-2i\theta} a_\theta) \geq 0$. Moreover, equality happens if and only if f and f_θ differ by a constant.

Let us consider the problem of maximizing $\mathrm{Re}(e^{-2i\theta} a_1)$ over all functions in $S_{z_0,\theta}$. For simplicity, we assume that $z_0 = \infty$. The proof of the other case is almost identical.

First, this class is not empty since $f(z) = z$ belongs to this class. By the Cauchy formula, the functional $f \mapsto a_1$ is continuous with respect to the local uniform convergence in $\widehat{\mathbb{C}}$. Since the family S_θ is normal, the normalization near z_0 implies that the limit of any convergent sequence is univalent. This follows from essentially the same argument as in the simply connected case. This gives us that there is a function $F(z) = z + A_1/z + \cdots$ which maximizes our functional over all functions from S_θ. It remains to be shown that this F maps onto a slit domain.

Assume that this is not true and there is a boundary component γ such that its image is not a slit with the slope θ. Let Ω' be the domain exterior to $F(\gamma)$. This domain is simply connected. By the simply connected Riemann mapping theorem, there is a conformal map from this domain onto the complement of the unit disc which maps infinity to itself. Composing this map with an appropriate Möbius transformation, a rotation of the Joukowsky map, we can find a map ϕ from Ω' onto a domain which is a complement of a slit with the slope θ. Moreover, we can choose this map to have expansion at infinity of the form $\phi(z) = z + a_\theta z^{-1} + \cdots$. We already know that $\operatorname{Re}\left(e^{-2i\theta}a_\theta\right) > 0$ since $F(\gamma)$ is not a straight slit. The composition $\phi(F)$ is also a univalent map on Ω. Expanding the composition at infinity, we get

$$\phi(F(z)) = z + \frac{a_\theta + A_1}{z} + \cdots.$$

This implies that $\phi(F) \in S_\theta$ and it has a larger value of the functional than F. This contradiction implies that F maps onto a slit domain.

Circular and Radial Slit Domains. These are two similar classes of slit domains consisting of the complex sphere without some slits. In the first case, we remove several arcs that lie on concentric circles centred at the origin. In the second case, we remove intervals that lie on rays emanating from the origin (see Figure 2.18).

In both cases, we can normalize a map in such a way that two given points z_1 and z_2 from Ω are mapped to the origin and infinity. Let us consider a family of functions f that are univalent in Ω, $f(z_1) = 0$, and there is a simple pole of residue 1 at z_2. The function that maximizes $|f'(z_1)|$ maps the domain onto a circular slit domain and the function that minimizes $|f'(z_1)|$ maps the domain onto a radial slit domain. We will not prove these results here. Interested readers can find the proofs and further discussion of uniformization of multiply connected domains in Nehari (1975, Chapter VII).

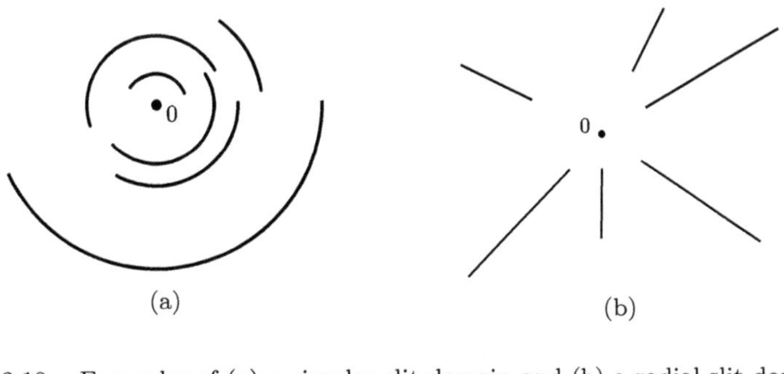

Fig. 2.18. Examples of (a) a circular slit domain and (b) a radial slit domain.

2.10. Solutions

Solution of 2.1. The simplest Möbius transformation from $\mathbb{H}$ onto $\mathbb{D}$ is $(z-i)/(z+i)$, and the inverse map is $i(1+z)/(1-z)$. Sometimes, it is convenient to have a Möbius transformation that maps $0 \in \partial\mathbb{H}$ to $1 \in \partial\mathbb{D}$. This could be done by

$$\begin{aligned} &\frac{i-z}{i+z}, \quad \mathbb{H} \to \mathbb{D}, \\ &i\frac{1-z}{1+z}, \quad \mathbb{D} \to \mathbb{H}. \end{aligned} \tag{2.18}$$

□

Solution of 2.2. First, we recall that any Möbius transformation is a conformal automorphism of the entire $\widehat{\mathbb{C}}$. This implies that if a Möbius transformation μ maps the boundary of some simply connected domain Ω onto some Jordan curve γ, then μ maps Ω onto one of the two components of $\widehat{\mathbb{C}} \setminus \gamma$. In other words, if we know the image of the boundary, then we almost know the image of the entire domain: we have only two possibilities. By considering the image of any point from Ω, we can figure out which of the two components is the image of Ω. It is also important to recall that a Möbius transformation maps circles and lines to circles and lines.

$\mathbb{D}$. First, we show that indeed all maps of the form

$$\mu(z) = e^{i\theta}\frac{z-a}{1-\bar{a}z}, \qquad a \in \mathbb{D}, \theta \in \mathbb{R}$$

map $\mathbb{D}$ to itself. It is easy to see that

$$|\mu(z)|^2 = \frac{z-a}{1-\bar{a}z}\frac{\bar{z}-\bar{a}}{1-\bar{z}a} = \frac{|z|^2 - a\bar{z} - \bar{a}z + |a|^2}{1 - a\bar{z} - \bar{a}z + |a|^2|z|^2}.$$

If $z \in \partial\mathbb{D}$, that is, $|z| = 1$, the last expression is equal to 1. This proves that the image of $\mathbb{D}$ is either $\mathbb{D}$ or $\mathbb{D}_-$. Since $\mu(a) = 0$, the image must be $\mathbb{D}$.

Next, let us assume that μ is a Möbius transformation from $\mathbb{D}$ to itself. Define $a = \mu(0)$ and $\phi(z) = (z-a)/(1-z\bar{a})$. From the above argument, we know that $\psi = \phi(\mu)$ is a Möbius transformation preserving the unit disc and the origin. We claim that ψ is a rotation, that is, $\psi(z) = e^{i\theta}z$ for some real θ. Explicit computation shows that in this case,

$$\mu(z) = e^{i\theta}\frac{z + ae^{-i\theta}}{1 + z\bar{a}e^{i\theta}}.$$

To prove the claim, we could use Lemma 2.2, which would definitely be overkill. Alternatively, we know that $\psi(z) = bz/(cz+d)$ with $b \neq 0$. By our assumption, $|\psi(e^{i\theta})| = 1$ for every real θ. Plugging in $e^{i\theta}$ and $-e^{i\theta}$, we find that $e^{i\theta}c$ and d must be orthogonal for every θ; hence $c = 0$ and $\psi(z) = bz/d$ with $|b/d| = 1$. This is equivalent to our claim.

$\mathbb{C}$. The general form of a Möbius transformation is

$$\frac{az+b}{cz+d}.$$

We know that this is a bijection of $\widehat{\mathbb{C}}$ onto itself. If it also maps $\mathbb{C}$ onto itself, it must map infinity to infinity. This happens if and only if $c = 0$.

$\mathbb{H}$. It is possible to prove this by composing Möbius automorphisms of the unit disc with Möbius maps (2.18), but it is easier to prove it directly. The Möbius transformation preserving $\mathbb{H}$ must be real on the real line. By considering the real values of z, it is a straightforward check that a, b, c, d must have the same argument and hence could be taken as real. To check the last condition, we consider

$$\mu(i) = \frac{ia+b}{ic+d} = \frac{ac+bd+i(ad-bc)}{c^2+d^2}.$$

This is in $\mathbb{H}$ if and only if $ad - bc > 0$. □

Solution of 2.3. All proofs are similar to the argument for the unit disc. Given a conformal automorphism, we compose it with a Möbius transformation that maps some of the points back and then argue that the composition is identity and hence the given transformation is an inverse of a Möbius map.

$\widehat{\mathbb{C}}$. Let f be a conformal automorphism of $\widehat{\mathbb{C}}$. Then there is a unique Möbius transformation μ such that $\mu(f(0)) = 0$, $\mu(f(1)) = 1$, and $\mu(f(\infty)) = \infty$. Let us denote this composition by g. We can write g as a Taylor series at the origin. Since $g(0) = 0$, there is no constant term and since the function is analytic at infinity, there are no terms of order more than one. This means that $g(z) = a_1 z$. From the last condition $g(1) = 1$, we get $g(z) = z$. This proves that f is Möbius.

$\mathbb{C}$. It is easy to see that $f(z) \to \infty$ as $z \to \infty$. By defining $f(\infty) = \infty$, any conformal automorphism of $\mathbb{C}$ can be extended to a conformal automorphism of $\widehat{\mathbb{C}}$ and we can apply the previous argument.

$\mathbb{H}$. There are many Möbius transformations mapping $\mathbb{D}$ to $\mathbb{H}$. The simplest example is $\mu(z) = i(1-z)/(z+1)$. If f is a conformal automorphism of $\mathbb{H}$, then $\mu^{-1} \circ f \circ \mu$ is a conformal automorphism of $\mathbb{D}$, and hence is a Möbius transformation. From this, we see that f is a composition of three Möbius transformations, which means that f is a Möbius transformation. □

Solution of 2.4. Let K be any compact subset of $\mathbb{D}$. Then there is $r < 1$ such that K is inside the disc of radius r and $|f_n(z)| = |z|^n \leq r^n$. This proves that $f_n \to 0$ uniformly on K.

If a domain Ω contains a point ζ with $|\zeta| = 1$, then there is a compact subset of Ω containing points with modulus 1 and smaller than 1, and hence the pointwise limit of f_n on this set is discontinuous and cannot be uniform. □

Solution of 2.5. Let $f_n = f_{w_n}$ be a sequence of functions. If the sequence $w_n \to \infty$, then $|f_{w_n}(z)| \leq 1/(|w_n| - 1) \to 0$, so the sequence converges uniformly to 0. Otherwise, there is a subsequence $w_{n_k} \to w$. We claim that f_{n_k} converge to f_w uniformly on compacts.

Let $K \subset \mathbb{D}$ be a compact. Then there is $r < 1$ such that K lies inside the disc of radius r. On this compact,

$$|f_{n_k}(z) - f_w(z)| = |w_{n_k} - w| \frac{|z|}{|z - w||z - w_{n_k}|} \leq |w_{n_k} - w| \frac{r}{|1 - r|^2}.$$

This proves the uniform convergence on K. □

Solution of 2.6. Let us consider $g(z) = (z-1)/(z+1)$, which is a Möbius transformation from the right half-plane onto the unit disc, and consider the family of functions $g(f)$ where $f \in \mathcal{F}$. This is a family of uniformly bounded analytic functions; hence, by Montel's theorem, it is normal.

Let f_n be a sequence in $\mathcal{F}$. Then from the sequence $g(f_n)$ we can find a subsequence $g(f_{n_k})$ which converges uniformly on every compact set. One can easily see that either the limit is identically equal to 1 or it is never equal to 1. Indeed, if $g = 1$, then it means that $|g|$ attains its maximal values, which means that g is identically equal to 1. If g is equal to 1, then f_{n_k} converge to infinity, which contradicts the assumption that $f_n(0)$ is bounded. If $g \neq 1$, since g^{-1} is uniformly continuous on compacts, the sequence $g^{-1}(g(f_{n_k})) = f_{n_k}$ also converges uniformly on compacts. This proves that $\mathcal{F}$ is normal. □

Solution of 2.7. We show that functions from our family are uniformly bounded on every disc $\mathbb{D}_r = \{|z| \leq r\}$. By Montel's theorem, this implies that the family is normal.

Let us choose some R such that $r < R < 1$ and let f be any function with $\|f\|_{H^1} \leq 1$. By the Cauchy formula, for any $z \in \mathbb{D}_r$,

$$f(z) = \frac{1}{2\pi i} \int_{|w|=R} \frac{f(w)}{w - z} \mathrm{d}z.$$

By the triangle inequality,

$$|f(z)| \leq \frac{1}{2\pi}\int_0^{2\pi} \frac{|f(w)|}{|z-w|} R d\theta \leq \frac{1}{R-r}\|f\|_{H^1} \leq \frac{1}{R-r}.$$

□

Solution of 2.8. Strip. The simplest map onto $\mathbb{H}$ is $\exp(\pi z)$. It maps the real line onto the positive real line and the line $\operatorname{Im} z = 1$ onto the negative real line. To map onto the unit disc, we compose with $(z-i)/(z+i)$, which is a Möbius transformation from $\mathbb{H}$ onto $\mathbb{D}$.

Two circles. As always, when working with domains whose boundary is made of circles, it is beneficial to apply a Möbius transformation that sends the circles to lines. In this case, we start with $\mu(z) = 1/z$. This transformation maps Ω onto an infinite strip $1/R < \operatorname{Re}(z) < 1/r$. This could be proved by either direct computation or by noting that the images of two circles must be parallel lines passing through $1/R$ and $1/r$. By symmetry with respect to the real line, they must be vertical lines. By applying $i(z - 1/r)/(1/R - 1/r)$, we map our strip to the strip $0 < \operatorname{Im} z < 1$ and reduce our problem to the previous one.

Half-plane with a slit. The simplest way is to note that z^2 maps Ω onto a particularly simple domain: $\mathbb{C} \setminus [-t^2, +\infty)$, adding t^2 and applying the appropriate branch of $z^{1/2}$ we map onto $\mathbb{H}$. In other words, the desired map is $\sqrt{z^2+t^2}$ (see Figure 2.19). Note that we use a non-standard version

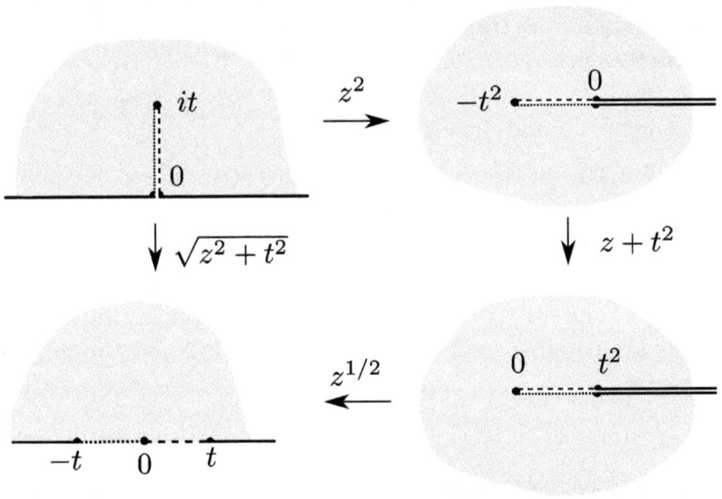

Fig. 2.19. The domain is the upper half-plane with a slit. The map $\sqrt{z^2+t^2}$ is not continuous on the slit. One side of the slit is dashed and the other is dotted; their images are $[0,t]$ and $[-t,0]$, respectively.

of the square root; in most cases, $\sqrt{z}$ denotes the branch of $z^{1/2}$ which has the positive real part, which corresponds to the branch cut along $(-\infty, 0]$. We use the branch which has the positive imaginary part (it corresponds to the branch cut $[0, +\infty)$). We will use this branch throughout this book.

Disc with a slit. First, we note that the problem for different values of x is essentially the same. Indeed, the Möbius map $(z-x)/(1-zx)$ maps $\mathbb{D} \setminus [x, 1]$ onto $\mathbb{D} \setminus [0, 1]$. Applying one more Möbius transformation $i(1-z)/(z+1)$, we map the domain onto $\mathbb{H}\setminus[0, i]$ and we can apply $\sqrt{z^2+1}$ from the previous part.

For positive x, it is convenient to have a map onto $\mathbb{D}$ which preserves the origin. This map will be useful later on, so we prefer to write it explicitly. The map that we have constructed above maps 0 to $2i\sqrt{x}/(1-x)$. Rescaling by $(1-x)/2\sqrt{x}$ and composing with $(i-z)/(i+z)$, we obtain a map from the disc with a slit onto the unit disc; moreover, this map preserves the origin. Combining all these steps and simplifying the resulting algebraic expression, we can write this map as

$$\frac{i\sqrt{x}(1+z) - \sqrt{(z-x)(1-xz)}}{i\sqrt{x}(1+z) + \sqrt{(z-x)(1-xz)}}. \tag{2.19}$$

□

Solution of 2.9. Let us assume that there are two functions like this: f_1 and f_2. By composing f_1^{-1} with f_2, we obtain a conformal map g from $\mathbb{H}$ onto itself such that $g(0) = 0$, $g(\infty) = \infty$, and $g'(0) = 1$. We know that this must be a Möbius transformation, that is, $g(z) = (az+b)/(cz+d)$ with real coefficients a, b, c, and d. From $g(0) = 0$, we have $b = 0$ and from $g(\infty) = \infty$, we have $c = 0$. Finally, from $g'(0) = 1$, we have $a/d = 1$, which means that $g(z) = z$ and $f_1 = f_2$. □

Solution of 2.10. Strip. This semi-infinite strip is a polygonal domain with three vertices: $-\pi/2$, $\pi/2$, and ∞. We can describe the map from the upper half-plane $\mathbb{H}$ onto S by the Christoffel–Schwarz formula. Since we have only three points, we can choose them arbitrary, and the easiest choice is -1, 1, and ∞ (which does not appear explicitly in the formula). Both angles are equal to $\pi/2$; hence $\beta_1 = \beta_2 = 1/2$ and the map is given by the integral

$$F(w) = \int_0^w (w-1)^{-1/2}(w+1)^{-1/2}\mathrm{d}w = \int_0^w \frac{1}{\sqrt{1-w^2}}\mathrm{d}w = \arcsin(w)$$

(we choose branches of the square root in such a way that the integrand is positive on $(-1, 1)$). Note that this function maps ± 1 to $\pm\pi/2$, so we do not need extra scaling or shifting. Another comment is that one can just note that $\sin(z)$ maps the boundary of S onto the real line and use this to prove that it maps S onto $\mathbb{H}$.

Triangle. As before, we have freedom to choose the points w_i in the Christoffel–Schwarz formula and we make the same choice. The only difference is that now $\alpha_i = 1/3$ and $\beta_i = 2/3$. The integral is

$$F(w) = \int_0^w (1-w)^{-2/3}(w+1)^{-2/3}\mathrm{d}w = \int_0^w \frac{1}{(1-w^2)^{2/3}}\mathrm{d}w.$$

The integrand is real on $(-1,1)$; hence this interval is mapped onto $(F(-1), F(1))$. The length of this interval is equal to

$$\begin{aligned} C &= \int_{-1}^{1} \frac{1}{(1-x^2)^{2/3}}\mathrm{d}x \\ &= \int_0^1 (1-t)^{1/3-1}t^{1/2-1}\mathrm{d}t \\ &= B\left(\frac{1}{2}, \frac{1}{3}\right), \end{aligned}$$

where B is the beta function. This proves that the map

$$C^{-1}F(w) + \frac{1}{2}$$

maps the upper half-plane onto an equilateral triangle which lies in the upper half-plane with side length 1 and vertices at 0 and 1 (see Figure 2.20).

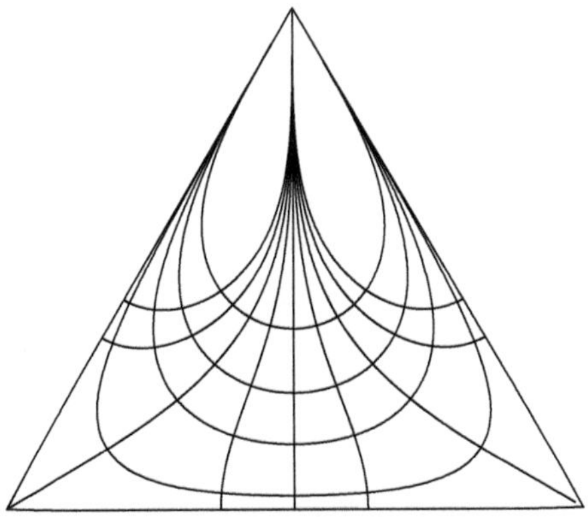

Fig. 2.20. The image of $\mathbb{H}$ under the Christoffel–Schwarz map. The lines inside the triangle are the images of horizontal and vertical lines.

Note that the integral in the definition of F can also be written in terms of special functions

$$F(w) = w\,{}_2F_1\left(\frac{1}{2}, \frac{2}{3}, \frac{3}{2}, w^2\right),$$

where ${}_2F_1(a, b, c, z)$ is the hypergeometric function defined as

$${}_2F_1(a, b, c, z) = \sum_{n=0}^{\infty} \frac{(a)_n (b)_n}{(c)_n\, n!} z^n,$$

and $(x)_n = x(x+1)\cdots(x+n-1)$ is the Pochhammer symbol. □

Solution of 2.11. Let us consider a side of R and the side of R' which is its image under f. Consider two Christoffel–Schwarz maps ϕ and ϕ' from $\mathbb{H}$ onto R and R' such that they send $[-1, 1]$ to these two sides. The preimages of other vertices are $\pm 1/k$ and $\pm 1/k'$ for R and R', respectively. Existence of the map f means that there is a Möbius transformation μ of $\mathbb{H}$ which maps $(-1/k, -1, 1, 1/k)$ to $(-1/k', -1, 1, 1/k')$. Since Möbius transformations preserve cross-ratios, $(k+1)^2/4k = (k'+1)^2/4k'$. Since both k and k' are in $(0, 1)$, this implies that $k = k'$ and the shapes of the rectangles are the same. □

Solution of 2.12. Let $(\tilde{\gamma}_n)$ be another null chain equivalent to (γ_n) and let $\tilde{D}_n$ be their insides. By the definition of equivalence, $\tilde{D}_n \subset D_m$ for every m and for all sufficiently large n. This proves that the impression defined in terms of $\tilde{D}_n$ is inside the impression defined by D_n. By symmetry, the other inclusion also holds and both impressions are equal. □

Solution of 2.13. From formula (2.5), we can see that

$$0 \le P(\zeta, z) \le (1 - |z|^2)/(1 - |z|)^2 = (1 + |z|)/(1 - |z|).$$

Since f is continuous, it is bounded and hence the integral in the definition of u is convergent. Moreover, in any disc of radius $\{z : |z| \le r < 1\}$, we have a uniform bound on P and its derivatives. This implies that it is possible to switch the order of differentiation and integration, and obtain

$$\Delta u(z) = \frac{1}{2\pi} \int f(\zeta) \Delta P(\zeta, z) |d\zeta| = 0.$$

The last equality follows from the fact that P is harmonic with respect to z. Since this works for any $r < 1$, the function u is harmonic in the entire unit disc.

From the construction of the Poisson kernel, or by an explicit computation, one can see that

$$\frac{1}{2\pi}\int P(e^{i\theta}, z)\mathrm{d}\theta = 1$$

and

$$u(z) - f(e^{i\theta_0}) = \frac{1}{2\pi}\int P(e^{i\theta}, z)\left(f(e^{i\theta}) - f(e^{i\theta_0})\right)\mathrm{d}\theta.$$

For z close to $\zeta_0 = e^{i\theta_0}$ this integral should be small. Indeed, when θ is far from θ_0, the Poisson kernel is small, and for θ close to θ_0, the second factor is small. To formalize this argument, we use that f is continuous at ζ_0 and for every $\varepsilon > 0$ there is $\delta > 0$ such that $|f(e^{i\theta}) - f(e^{i\theta_0})| < \varepsilon$ for $|\theta - \theta_0| < \delta$.

We assume that $|\theta - \theta_0| > \delta$ and that $|z - e^{i\theta_0}| < \sigma < \delta$. From the formula for the Poisson kernel, we see that in this case,

$$P(e^{i\theta}, z) \leq \frac{2\sigma}{(\delta - \sigma)^2} < \varepsilon,$$

where the last inequality holds for sufficiently small $\sigma = \sigma(\delta)$. Now, for an arbitrarily small $\varepsilon > 0$, we choose δ and then σ, so that the above inequalities hold. For this choice of constants, we have

$$\begin{aligned}|u(z) - f(e^{i\theta_0})| &\leq \frac{1}{2\pi}\int_{|\theta-\theta_0|<\delta} P(e^{i\theta}, z)|f(e^{i\theta}) - f(e^{i\theta_0})|\mathrm{d}\theta \\ &\quad + \frac{1}{2\pi}\int_{|\theta-\theta_0|\geq\delta} P(e^{i\theta}, z)|f(e^{i\theta}) - f(e^{i\theta_0})|\mathrm{d}\theta \\ &\leq \frac{1}{2\pi}\left(\varepsilon\int P(e^{i\theta}, z)\mathrm{d}\theta + 2M\varepsilon\int \mathrm{d}\theta\right) = \varepsilon\left(\frac{1}{2\pi} + 2M\right),\end{aligned}$$

where M is the supremum of f. This proves that $u(z) \to f(\zeta)$ as $z \to \zeta$. □

Solution of 2.14. Let $z \in \mathbb{H}$ and consider $R > |z|$. Consider a contour Γ_R which is made of the interval $[-R, R]$ and the upper semi-circle $Re^{i\theta}$, $\theta \in [0, \pi]$. By the Cauchy formula,

$$f(z) = \frac{1}{2\pi i}\int_{\Gamma_R} \frac{f(w)}{w - z}\mathrm{d}w.$$

As $R \to \infty$, and since f vanishes at infinity, the integral over the semi-circle tends to zero. Hence, passing to the limit, we get

$$f(z) = \frac{1}{2\pi i} \int_{-\infty}^{\infty} \frac{f(t)}{t - z} \mathrm{d}t.$$

Using the same argument and the fact that $\bar{z}$ is outside of the contour, we have

$$\frac{1}{2\pi i} \int_{-\infty}^{\infty} \frac{f(t)}{t - \bar{z}} \mathrm{d}t = 0.$$

Considering the complex conjugate of the last integral and using that on the real line $\bar{t} = t$, we have

$$\int_{-\infty}^{\infty} \frac{\bar{f}(t)}{t - z} \mathrm{d}t = 0.$$

Reintroducing the factor of $1/2\pi i$ and subtracting it from the formula for $f(z)$, we get

$$f(z) = \frac{1}{2\pi i} \int_{-\infty}^{\infty} \frac{f(t) - \bar{f}(t)}{t - z} \mathrm{d}t = \frac{1}{\pi} \int_{-\infty}^{\infty} \frac{\mathrm{Im}(f(t))}{t - z} \mathrm{d}t.$$

To obtain the real version, we add integrals instead of subtracting them. □

Solution of 2.15. As mentioned before, we immediately have that $G(z) = G_{\Omega'}(f(z_0), f(z))$ is harmonic in Ω and vanishes on the boundary.

By definition, we have $G(z) = g(f(z)) - \log|f(z) - f(z_0)|$, where g is a bounded harmonic function. Let us consider

$$\log|f(z) - f(z_0)| = \log\left|\frac{f(z) - f(z_0)}{z - z_0}\right| + \log|z - z_0|.$$

The function $(f(z) - f(z_0))/(z - z_0)$ is analytic and non-zero outside of z_0. At z_0, it has a removable singularity since f is differentiable. This implies that the first term could be extended by continuity to a bounded harmonic function.

Combining everything together, we have

$$G(z) = \left(g(f(z)) - \log\left|\frac{f(z) - f(z_0)}{z - z_0}\right|\right) - \log|z - z_0|,$$

where the terms in parenthesis are bounded harmonic functions. This proves that $G(z) = G_{\Omega}(z_0, z)$. □

Solution of 2.16. It is possible to derive the formula for the Poisson kernel directly from the second definition of the harmonic measure, but we prefer to derive it from the first definition.

Let us consider

$$f(w) = \frac{w - z}{w - \bar{z}}.$$

This is a Möbius transformation which maps $\mathbb{H}$ onto $\mathbb{D}$ and z to 0. By conformal invariance, the harmonic measure in $\mathbb{H}$ is the pull back of the harmonic measure in $\mathbb{D}$. The density of harmonic measure is $1/2\pi$. The map f is analytic on the boundary of the domain; hence the density of harmonic measure is changed by the derivative of the map. Simple computation gives us that $|f'(t)| = 2y/((x - t)^2 + y^2)$ for real t and hence the density is

$$\frac{1}{\pi} \frac{y}{y^2 + (x - t)^2}. \qquad \square$$

Solution of 2.17. First, without loss of generality, we can assume that $r_1 = r_2 = 1$ and that f maps the unit circle to itself. For a circle of radius r centred at the origin, we define the inversion by $I_r(z) = r^2/\bar{z}$. We can extend f to the annulus $A(R_1, R_1^2)$ by $f(z) = I_{R_2}(f(I_{R_1}))$. By the reflection principle, this gives us a conformal map from $A(1, R_1^2)$ onto $A(1, R_2^2)$ which is also continuous up to the boundary.

Repeating the same argument, we can extend f to a conformal map from the annulus $A(R_1^{-n}, R_1^n)$ onto the annulus $A(R_2^{-n}, R_2^n)$ which is continuous up to the boundary, and for each integer k with $|k| \leq n$, it maps the circle $\{|z| = R_1^k\}$ onto the circle $\{|z| = R_2^k\}$. Passing to the limit as n tends to infinity, we have a map from $\mathbb{C} \setminus \{0\}$ onto itself. Defining $f(0) = 0$, we have a conformal map from $\mathbb{C}$ to itself. This map must be linear, but since $f(0) = 0$ and the unit circle is mapped onto itself, it must be of the form $f(z) = az$ where $|a| = 1$, that is, f is a rotation. This immediately implies that $R_1 = R_2$. $\square$

Solution of 2.18. By applying a shift and rotation, we can assume that the outer circle is $\{|z| = 1\}$ and the centre of the inner circle is on the real line. Let us denote the points where the inner circle intersects the real line by x_1 and x_2. If $x_1 = -x_2$, then the circles are concentric and there is nothing to do. Hence, we can safely assume that $x_1 \neq -x_2$.

We are looking for a Möbius transformation μ which will preserve the unit disc and the upper half-plane (hence the real line), such that $\mu(x_1) = -\mu(x_2)$. By symmetry and properties of Möbius transformations, this function will preserve the outer circle and map the inner circle to a circle centred at the origin. Hence it will map our domain to an annulus.

From the property of Möbius transformations preserving the real line and preserving $\mathbb{D}$, we have that μ must be of the form $\mu(z) = (z - a)(1 - az)$, where $a \in (-1, 1)$. The equation $\mu(x_1) = -\mu(x_2)$ is quadratic in a and can be solved explicitly. It has two roots, which multiply to 1, and hence only one of them is from $(-1, 1)$. Direct computations give us that

$$a = \frac{1 - x_1 x_2 - \sqrt{(1 - x_1^2)(1 - x_2^2)}}{x_1 + x_2}. \qquad \square$$

Chapter 3

Basic Theory of Univalent Maps

In this chapter, we discuss general properties of univalent functions. We pay special attention to universal boundary estimates and connections between analytic and geometric properties.

3.1. Classes S and Σ

One of the most important classes of univalent functions is the class S (from the German word *schlicht*, which is another standard term for univalent functions) consisting of all univalent functions in the unit disc normalized by conditions $f(0) = 0$ and $f'(0) = 1$. Alternatively, they are given by the Taylor series of the form

$$f(z) = z + a_2 z^2 + a_3 z^3 + \cdots$$

that converge in the unit disc.

For any non-trivial simply connected domain, there is a univalent function from $\mathbb{D}$ onto this domain. By rescaling and shifting the domain, the function can be normalized to be from the class S. So, up to scaling and translations, functions from S describe all simply connected domains except, of course, $\mathbb{C}$ and $\widehat{\mathbb{C}}$.

Another standard class is the family of functions that are univalent in the complement of the unit disc $\mathbb{D}_-$ and have the expansion

$$g(z) = z + b_0 + b_1 z^{-1} + b_2 z^{-2} + \cdots .$$

We denote this class by Σ. Each $g \in \Sigma$ maps $\mathbb{D}_-$ onto the complement of a compact set E. Sometimes, it is more convenient to assume that $0 \in E$.

This subclass of Σ is denoted by Σ'. Note that any function from Σ differs from some function from Σ' by subtraction of an appropriate constant, so these two classes are extremely close and share most of their properties.

One of the reasons for introduction of Σ' is that there is a simple bijection between functions from S and Σ'. If f is an arbitrary function from the class S, then

$$g(z) = \frac{1}{f(1/z)}$$

belongs to the class Σ'. Conversely, for every $g \in \Sigma'$,

$$f(z) = \frac{1}{g(1/z)} \in S.$$

For functions given by the Taylor series, it is generally very difficult to check whether they are in S or not. There are some sufficient conditions but they are rather weak and cover only special cases.

A very important example of a function from S is the *Koebe function*

$$K(z) = z + 2z^2 + 3z^3 + 4z^4 + \cdots = \sum_{n=1}^{\infty} nz^n.$$

It is difficult to see that $K \in S$ just by looking at the Taylor series. Fortunately, it is rather simple to write this series in a closed form as $z/(1-z)^2$. There are two standard ways to show that K is univalent. The first is to observe that it is a rational function of degree 2 and hence it is two-to-one on the complex sphere. By explicit computations, one can show that the unit circle is mapped onto $[-\infty, -1/4]$ and for all points outside of this ray only one preimage is inside the unit disc.

An alternative and more instructive way is to rewrite K as

$$\frac{1}{4}\left(\frac{1+z}{1-z}\right)^2 - \frac{1}{4}.$$

We know that $(1+z)/(1-z)$ is a Möbius transformation from the unit disc onto the right half-plane $\{z : \operatorname{Re} z > 0\}$. The square function maps it conformally onto the plane with a cut along the negative real line; scaling and subtracting $1/4$ maps it onto the plane with cut from $-\infty$ to $-1/4$. Since all maps here are obviously univalent, their composition is also univalent.

Exercise 3.1. Show that $K_\alpha \in S$ and find $K_\alpha(\mathbb{D})$, where

$$K_\alpha(z) = \frac{1}{2\alpha}\left[\left(\frac{1+z}{1-z}\right)^\alpha - 1\right], \quad \alpha \in (0,2].$$

Fig. 3.1. Joukowsky airfoil: image of a circle passing through 1 under the map $J(z)$. Here, we show the image of the circle centred at $0.2 + 0.2i$.

Exercise 3.2. Show that the Joukowsky function $J = z + 1/z$ belongs to Σ and find $J(\mathbb{D}_-)$ (Figure 3.1 shows the image of another circle). Show that the modified Joukowsky function $J_k(z) = z + k/z$ is also in Σ for all $-1 < k < 1$. Find the image $J_k(\mathbb{D}_-)$.

Exercise 3.3. Let f be a function from class S. Prove that the following functions are also from S:

(1) Let μ be a Möbius transformation preserving $\mathbb{D}$, then we can define

$$f_\mu(z) = \frac{f(\mu(z)) - f(\mu(0))}{(f(\mu))'(0)}. \tag{3.1}$$

When $\mu(z) = (z + w)/(1 + z\bar{w})$, this transformation is called the *Koebe transformation of f with respect to w*. When $\mu(z) = e^{i\theta}z$, the transformation becomes $f_\theta(z) = e^{-i\theta}f(e^{i\theta}z)$ and is called the *rotation of f*.

(2) Reflection of f is defined as $\bar{f}(\bar{z})$.

(3) *Koebe* or *root transform* is defined by

$$K_n(f)(z) = f^{1/n}(z^n) \tag{3.2}$$

(we also have to show that $K_n f$ can be defined as a single-valued function for all positive integers n). The same is true for functions from the class Σ'.

3.2. Bieberbach–Koebe Theory

The first example of a theorem relating to analytical properties with geometrical aspects is Gronwall's theorem, which relates the area of the complementary domain E with coefficients of a function $f \in \Sigma$.

Theorem 3.1 (Gronwall's area theorem). *Let $g(z) = z + \sum b_n z^{-n}$ be a function from the class Σ which maps $\mathbb{D}_-$ onto the complement of a compact set E. The area of E is given by*

$$m(E) = \pi\left(1 - \sum_{n=1}^{\infty} n|b_n|^2\right).$$

The proof of this theorem uses essentially the same technique as in the proof of Proposition 2.50.

Proof of Theorem 3.1. To compute the area of E, we would like to use Green's theorem for the image of the unit circle. This does not work since the function is not defined on the unit circle and it might be that it cannot even be continuously extended to the boundary. Instead, we use one of the standard tricks. Take some $r > 1$ and denote by γ_r the image of the circle $|z| = r$ under g. Since g is a univalent map, we have that γ_r is a simple closed analytic curve enclosing $E_r \supset E$. By using Green's formula in its complex form (1.2), the area of E_r is

$$\begin{aligned} m(E_r) &= \frac{1}{2i}\int_{\gamma_r} \bar{w}\mathrm{d}w = \frac{1}{2i}\int_{|z|=r} \bar{g}(z)g'(z)\mathrm{d}z \\ &= \frac{1}{2i}\int_0^{2\pi} \left(\bar{z} + \sum \bar{b}_n\bar{z}^{-n}\right)\left(1 - \sum nb_n z^{-n-1}\right) rie^{i\theta}\mathrm{d}\theta \\ &= \pi\left(r^2 - \sum_{n=1}^{\infty} n|b_n|^2 r^{-2n}\right). \end{aligned}$$

Passing to the limit as $r \to 1$, we complete the proof the theorem. □

Corollary 3.2. *Since the area of E is non-negative, we have*

$$\sum_{n=1}^{\infty} n|b_n|^2 \leq 1,$$

and in particular,

$$|b_n| \leq \frac{1}{\sqrt{n}}.$$

This inequality is sharp for $n = 1$ since $J(z) = z + 1/z$ is univalent, but not sharp for $n \geq 2$. If $|b_n| = 1/\sqrt{n}$, then $g(z) = z + b_0 + e^{i\theta}z^{-n}/\sqrt{n}$. Direct computation shows that g' vanishes at some points in $\mathbb{D}_-$, and hence g is not univalent.

Using these inequalities on the coefficients of functions from Σ, one can estimate the second coefficient of a function from S. This theorem was proved by Bieberbach (1916) and turns out to be very useful and influential.

Theorem 3.3 (Bieberbach). *Let $f(z) = z + \sum_{n=2}^{\infty} a_n z^n$ be a function from S. Then $|a_2| \leq 2$. Moreover, $|a_2| = 2$ if and only if f is a rotation of the Koebe function.*

Proof. As we discussed before, the function $1/f(1/z)$ is from class Σ'. Applying the Koebe square root transform (see (3.2)) with $n = 2$, we see that

$$g(z) = \frac{1}{\sqrt{f(1/z^2)}}$$

is also from Σ'. From the Taylor series for f, we compute

$$\sqrt{f(z^2)} = \sqrt{z^2 + a_2 z^4 + \cdots} = z\sqrt{1 + a_2 z^2 + \cdots} = z\left(1 + \frac{a_2}{2}z^2 + \cdots\right)$$

and

$$g(z) = \frac{z}{1 + a_2 z^{-2}/2 + \cdots} = z - \frac{a_2}{2}z^{-1} + \cdots .$$

Applying Corollary 3.2, we get $|a_2|/2 \le 1$, with equality holding if and only if

$$g(z) = z - e^{i\theta} z^{-1}$$

for some real θ. Rewriting f in terms of g, we get

$$f(z) = e^{-i\theta}\left(\frac{e^{i\theta} z}{(e^{i\theta} z - 1)^2}\right) = e^{-i\theta} K(e^{i\theta} z). \qquad \square$$

The Bieberbach theorem implies a very important corollary about the geometrical properties of functions from S. By analyticity, we know that $\Omega = f(\mathbb{D})$ contains an open neighbourhood of the origin. The lower bound on the distance from the origin to the boundary of Ω is as follows.

Theorem 3.4 (Koebe 1/4 theorem). *Let f be a function from S. Then $\Omega = f(\mathbb{D})$ contains the disc $\{z : |z| < 1/4\}$. Moreover, if there is $w \notin \Omega$ with $|w| = 1/4$, then f is a rotation of the Koebe function.*

Proof. Let us take any point w which is not in Ω. Since $w \neq 0$, the function

$$\phi(z) = \frac{w f(z)}{w - f(z)}$$

is a composition of a Möbius transformation and f; hence it is univalent. Expanding everything into the power series, we get

$$\phi(z) = \frac{w f(z)}{w - f(z)} = \frac{wz + w a_2 z^2 + \cdots}{w - z - a_2 z^2 - \cdots} = z + \left(a_2 + \frac{1}{w}\right) z^2 + \cdots ,$$

which implies that $\phi \in S$ and, by the Bieberbach theorem, $|a_2 + 1/w| \leq 2$. Since $|a_2| \leq 2$, we have $|1/w| \leq 4$ or, equivalently, $|w| \geq 1/4$. This proves that all points with $|w| < 1/4$ must be in Ω.

To prove the last part, we note that $|w| = 1/4$ implies that $|a_2| = 2$ and the Bieberbach theorem (Theorem 3.3) implies that f must be a rotation of the Koebe function. □

If, on the other hand, $\mathbb{D} \subset \Omega$, then the Schwarz lemma (Theorem 2.2) applied to $f^{-1}|_{\mathbb{D}}$ implies that $f(z) = z$. This argument implies that Ω cannot contain a disc centred at the origin of radius larger than 1. Together with the Koebe 1/4 theorem, we have the following corollary.

Corollary 3.5. *Let $f : \mathbb{D} \to \Omega$ be a function from S. Then* $\mathrm{dist}(0, \partial\Omega) \in [1/4, 1]$.

By rescaling, we can apply the same result to general univalent functions in $\mathbb{D}$. Let $f : \mathbb{D} \to \Omega$ be a univalent map. Denote $f(0)$ by z_0. Considering $g(z) = (f(z) - z_0)/f'(0)$, which is from S, we have

$$1 \leq \frac{|f'(0)|}{\mathrm{dist}(z_0, \partial\Omega)} \leq 4.$$

Recall that $|f'(0)| = \rho(z_0, \Omega)$ is the conformal radius of Ω from z_0. Hence, the above inequality can be reformulated in somewhat more geometric terms. It is a statement that the conformal radius is comparable to the distance to the boundary (which is sometimes called the *inradius*).

Corollary 3.6. *Let Ω be a simply connected domain and z_0 be a point inside. Then*

$$1 \leq \frac{\rho(z_0, \Omega)}{\mathrm{dist}(z_0, \partial\Omega)} \leq 4.$$

The Koebe 1/4 theorem can be easily generalized to arbitrary univalent maps.

Theorem 3.7 (Koebe distortion theorem). *Let $f : \Omega \to \Omega'$ be a univalent map and let z be some point in Ω. Then*

$$\frac{1}{4}\mathrm{dist}(f(z), \partial\Omega') \leq |f'(z)|\mathrm{dist}(z, \partial\Omega) \leq 4\mathrm{dist}(f(z), \partial\Omega'). \qquad (3.3)$$

Since $f(z) - f(z_0) \approx f'(z_0)(z - z_0)$, the distances are distorted locally by $|f'|$. The Koebe theorem tells us that in some sense the same holds globally up to a constant which is between 1/4 and 4.

Exercise 3.4. Prove the Koebe distortion theorem.

Exercise 3.5. Let f be a bounded univalent function in $\mathbb{D}$. Prove that

$$\lim_{|z|\to 1}(1-|z|)|f'(z)| = 0.$$

Give an example showing that boundedness is an essential condition.

Exercise 3.6. Let f be a univalent function in $\mathbb{D}$. Show that for all $z \in \mathbb{D}$

$$\frac{1}{4}(1-|z|^2)|f'(z)| \leq \operatorname{dist}(f(z), \partial f(\mathbb{D})) \leq (1-|z|^2)|f'(z)|.$$

[*Hint: Use the Koebe transform with respect to* z; *see* (3.1).]

We conclude this section with the sharp bounds on the distortion (that is, on $|f'|$) and on the growth (that is, on $|f|$) near the boundary. Both results will follow from the following inequality, due to Bieberbach (1916), which in turn follows from the coefficient estimate.

Theorem 3.8 (Bieberbach inequality). *Let* f *be a function from* S, *and* z *be any point with* $|z| = r < 1$. *Then*

$$\left|\frac{zf''(z)}{f'(z)} - \frac{2r^2}{1-r^2}\right| \leq \frac{4r}{1-r^2}. \tag{3.4}$$

Moreover, this inequality is sharp.

Proof. For $w_0 \in \mathbb{D}$, we can define the function

$$\phi(z) = \frac{f\left(\frac{z+w_0}{1+\bar{w}_0 z}\right) - f(w_0)}{(1-|w_0|^2)f'(w_0)}, \tag{3.5}$$

which is the Koebe transformation of f with respect to w_0 (see Exercise 3.3). Straightforward but messy computation gives the second term in the Taylor series

$$\phi(z) = z + \left(\frac{1}{2}(1-|w_0|^2)\frac{f''(w_0)}{f'(w_0)} - \bar{w}_0\right) z^2 + \cdots. \tag{3.6}$$

By Theorem 3.3, the second coefficient is bounded by 2. Hence,

$$\left|\frac{1}{2}(1-|w_0|^2)\frac{f''(w_0)}{f'(w_0)} - \bar{w}_0\right| \leq 2.$$

Changing w_0 to z and multiplying by $2z/(1-|z|^2)$ we obtain (3.4).

Direct computations for the Koebe function $K(z)$ and $z = r$ show that the inequality is sharp. By rotating the Koebe function, we can see that it is sharp for all radial directions. □

In the inequality (3.4), we can change the modulus to the real or imaginary part and obtain

$$\frac{-4r+2r^2}{1-r^2} \le \operatorname{Re}\left(\frac{zf''(z)}{f'(z)}\right) \le \frac{4r+2r^2}{1-r^2} \tag{3.7}$$

and

$$\frac{-4r}{1-r^2} \le \operatorname{Im}\left(\frac{zf''(z)}{f'(z)}\right) \le \frac{4r}{1-r^2}.$$

But $zf''/f' = r\partial_r(\log f')$ and the inequalities above can be rewritten as

$$\frac{-4+2r}{1-r^2} \le \partial_r \log|f'(z)| \le \frac{4+2r}{1-r^2} \tag{3.8}$$

and

$$\frac{-4}{1-r^2} \le \partial_r \arg f'(z) \le \frac{4}{1-r^2}. \tag{3.9}$$

These two inequalities lead to the following two theorems giving universal bounds on the modulus and argument of f'.

Theorem 3.9 (Distortion theorem). *For a function $f \in S$ and $r = |z|$, we have*

$$\frac{1-r}{(1+r)^3} \le |f'(z)| \le \frac{1+r}{(1-r)^3}.$$

Moreover, these inequalities are sharp; if the equality occurs in either of them for some $z \neq 0$, then f must be a rotation of the Koebe function.

Proof. Inequalities are obtained by integrating (3.8) along the straight interval from 0 to z. To prove the last part, we note that for the equality to hold for some $z = re^{i\theta}$, there must be equality in (3.8) and (3.7) for all $z = te^{i\theta}$, $t \in [0, r]$. Dividing (3.7) by t and passing to the limit $t \to 0$, we have

$$\operatorname{Re}\left(e^{i\theta}\frac{f''(0)}{f'(0)}\right) = \pm 4,$$

which in turn implies that the second coefficient of f has modulus 2 which happens only for rotations of the Koebe function. This argument, or the direct computation of the derivative of the Koebe function, shows that the inequality is indeed sharp. □

Theorem 3.10 (Rotation theorem). *For a function $f \in S$ and $r = |z|$, we have*

$$|\arg f'(z)| \leq 2 \log \frac{1+r}{1-r}.$$

Proof. This inequality is obtained by integrating the upper bound in (3.9) along the straight interval from 0 to z. □

The Rotation theorem in this form was proved by Bieberbach in 1919. Unlike the Distortion theorem, it is not sharp. The sharp bound has an unexpected form: it looks different for different ranges of $|z|$:

$$|\arg f'(z)| \leq \begin{cases} 4 \arcsin(r), & |z| = r \leq 1/\sqrt{2}, \\ \pi + \log \dfrac{r^2}{1-r^2}, & |z| = r \geq 1/\sqrt{2}. \end{cases}$$

The sharp upper bound was obtained by Goluzin (1936). The proof of the sharp estimate uses Loewner evolution techniques (see Chapter 5). The proof is similar in spirit to the argument in Section 5.1.3 but more technical; we refer interested readers to Duren (1983, Theorem 3.7).

Finally, we prove the universal estimates on the modulus of functions from S.

Theorem 3.11 (Growth theorem). *For a function $f \in S$ and $r = |z|$, we have*

$$\frac{r}{(1+r)^2} \leq |f(z)| \leq \frac{r}{(1-r)^2}.$$

Moreover, if the equality occurs in either of these inequalities for some $z \neq 0$, then f is a rotation of the Koebe function.

Proof. The upper bound is a simple corollary of the Distortion theorem (Theorem 3.9). Indeed, for $z = re^{i\theta}$, we have

$$f(z) = \int_0^r f'(se^{i\theta}) e^{i\theta} \mathrm{d}s.$$

By the triangle inequality and Theorem 3.9,

$$|f(z)| \leq \int_0^r \frac{1+s}{(1-s)^3} \mathrm{d}s = \frac{r}{(1-r)^2}.$$

To get the lower bound, we fix r and observe that it is enough to prove the inequality for z such that $|f(z)|$ is minimal. Let us consider the curve in $\Omega = f(\mathbb{D})$ which is the image of the circle of radius r. This curve is a

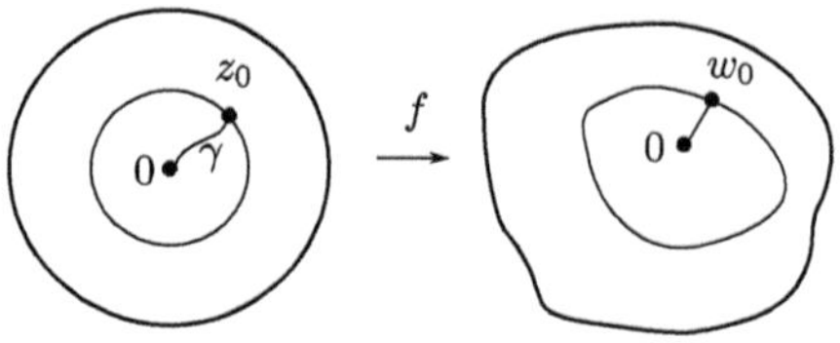

Fig. 3.2. The curve γ is constructed in such a way that the argument of $f'(z)\mathrm{d}z$ is constant on the curve.

compact set which does not contain 0. Let w_0 be a point on this curve which minimizes the distance to the origin. The interval from 0 to w_0 lies completely inside Ω. We denote its preimage by γ, which is obviously a simple curve connecting the origin with some point z_0 of modulus r and stays inside the closed disc or radius r (see Figure 3.2). By construction, $|f(z_0)| = \min_{|z|=r} |f(z)|$. As before, $f(z_0) = \int_\gamma f'(z)\mathrm{d}z$, but in this case our construction implies that the argument of $f'(z)\mathrm{d}z$ is constant along γ so we have

$$|f(z_0)| = \int_\gamma |f'(z)||\mathrm{d}z| \geq \int_0^r \frac{1-r}{(1+r)^3}\mathrm{d}r = \frac{r}{(1+r)^2}.$$

Since both inequalities are obtained by the integration of the inequalities from the Distortion theorem, the equality in any of them implies equality in the Distortion theorem, which in turn implies that the function is a rotation of the Koebe function. □

Exercise 3.7. Let $f \in S$ and $z \in \mathbb{D}$. By considering the Koebe transform, show that

$$\frac{1-r}{1+r} \leq \left| z\frac{f'(z)}{f(z)} \right| \leq \frac{1+r}{1-r}, \tag{3.10}$$

where $r = |z|$.

Exercise 3.8. Let f be a function from the class Σ'. Show that for every $|z| > 1$,

$$\frac{(1-|z|)^2}{|z|} \leq |f(z)| \leq \frac{(1+|z|)^2}{|z|}.$$

This inequality is also known as the Koebe distortion theorem.

Exercise 3.9. Let f_n be a family of conformal maps $f_n : \mathbb{D} \to \Omega_n$ such that $f_n(0) = 0$ and $f_n'(0) > 0$. Additionally, we assume that the domains are expanding to infinity, namely that $f_n'(0) \to \infty$ as $n \to \infty$. Equivalently, the conformal radius of Ω_n with respect to 0 tends to infinity. Define $g_n = f_n^{-1}$. Show that

$$|f_n'(0)|g_n(z) = \frac{g_n(z)}{|g_n'(0)|} \to z$$

uniformly on compacts.

Similarly, if $f_n : \mathbb{D}_- \to \widehat{\mathbb{C}}$ such that $f_n'(\infty) \to 0$ and $f_n(z)/|f_n'(\infty)|$ are from the class Σ', then the rescaled inverse maps

$$|f_n'(\infty)|g_t(z) \to z$$

uniformly on compact sets.

Combining the Growth theorem with the Montel theorem, we get an important corollary, which states that the class S is compact.

Corollary 3.12. *Let f_n be a sequence of functions in S. Then there is a subsequence f_{n_k} which converges uniformly on compacts to a function $f \in S$.*

Proof. Let K be a compact in $\mathbb{D}$; then there is $r < 1$ such that $|z| < r$ for all $z \in K$. Hence, on K, we have $|f| \le r/(1-r)^2$. This proves that S is uniformly bounded on compacts and by Montel's theorem S is normal. This proves that there is a subsequence f_{n_k} which converges uniformly on compacts to f.

We know that f is analytic; moreover, by the Hurwitz theorem, f is either univalent or constant. From the uniform convergence, we have that $f(0) = 0$ and $f'(0) = 1$, which proves that f cannot be constant, so it must be univalent; moreover, it has the right normalization, so it belongs to S. □

3.3. Sequences of Univalent Functions

As we discussed before, the Riemann mapping theorem gives us the correspondence between the simply connected domains and univalent maps in the unit disc. Given this correspondence, it is very natural to ask the question: Is convergence of univalent maps equivalent to convergence of domains? This is not a well-posed question since there are many different notions of convergence for both maps and domains.

It is easy to see that the answer might be 'no'. Let us consider $\Omega_n = n\mathbb{D}$ and $f_n : \mathbb{D} \to \Omega_n$ given by $f_n(z) = nz$. In this case, one would expect that any reasonable notion of convergence would give $\Omega_n \to \mathbb{C}$. On the other hand, $f_n(z)$ converges pointwise to infinity everywhere except the origin. Moreover, there are no conformal maps from $\mathbb{D}$ to $\mathbb{C}$, and thus there is no possible way for functions $\to \Omega_n$ to converge to a univalent map $f : \mathbb{D} \to \mathbb{C}$.

For convergence of maps, it is natural to use the standard convergence in the theory of univalent functions: uniform convergence on compact subsets. Convergence of domains is less obvious. In the usual geometric notions of convergence, it might happen that a limit of simply connected domains might not be simply connected or not even connected at all (see Example 3.15).

The right definition of convergence was introduced by Carathéodory (1912) using the notion of the kernel.

Definition 3.13. Let (Ω_n) be a family of simply connected domains in $\mathbb{C}$ containing a fixed point w_0. The *kernel* (with respect to w_0) of the family is the set of all points w in the plane such that there exists a simply connected domain D containing both w and w_0 and contained in all Ω_n for sufficiently large n. If there are no such points, then we define the kernel to be $\{w_0\}$.

We say that Ω_n converges to the kernel Ω if Ω is the kernel for every subsequence of (Ω_n). This convergence is called *Carathéodory* or *kernel* convergence.

There is an equivalent definition of the kernel which can be useful sometimes.

Definition 3.14. The kernel of (Ω_n) with respect to w_0 is the largest simply connected domain Ω such that $w_0 \in \Omega$ and every closed subset of Ω belongs to all Ω_n for sufficiently large n. If there are no such domains, then we define the kernel to be $\{w_0\}$.

Exercise 3.10. Show that the notions of kernel explained in Definitions 3.13 and 3.14 are equivalent.

The definitions naturally extend to families of domains $(\Omega_t)_{t\in[T_1,T_2]}$, the only difference is that 'for all sufficiently large n' should be replaced by 'for all t sufficiently close to T_2' if $T_2 < \infty$ and by 'for all sufficiently large t' if $T_2 = \infty$.

In some cases, Carathéodory convergence is the same as the topological one. It is easy to see that if the sequence Ω_n is increasing, that is, $\Omega_n \subset \Omega_{n+1}$, show that the kernel is $\Omega = \bigcup \Omega_n$ and $\Omega_n \to \Omega$ in the sense of Carathéodory.

In general, this is not the case, even for decreasing domains. The following example also shows that the kernel indeed depends on the reference point.

Example 3.15. Let us consider the following sequence of domains: $\Omega_n = \mathbb{C} \setminus A_n$, where $A_n = \{e^{i\theta},\ \theta \in [1/n, 2\pi - 1/n]\}$. If $w_0 \in \mathbb{D}$, show that the kernel is $\mathbb{D}$, if $w_0 = 1$, then the kernel is $\{1\}$, and, finally, if w_0 is outside of $\mathbb{D}$, then the kernel is $\mathbb{D}_-$. In all of these cases, Ω_n converges to the kernel. One might assume that the limit is the part of the domains seen from w_0.

For decreasing domains, the kernel might be different from the topological limit, but it is still rather easy to find.

Exercise 3.11. Let $\Omega_{n+1} \subset \Omega_n$ be a decreasing sequence of simply connected domains and let $\Omega = \bigcap \Omega_n$ be its topological limit. If w_0 is in the interior of Ω, show that the kernel with respect to w_0 is the component of interior of Ω which contains w_0, otherwise the kernel is $\{w_0\}$.

Now we can formulate the main theorem about the convergence of the univalent maps. For simplicity of notation, the theorem is stated for sequences of functions and domains. It is easy to see that the same proof works for families of functions and domains.

Theorem 3.16 (Carathéodory Convergence). *Let (Ω_n) be a sequence of simply connected domains containing w_0. Let f_n be conformal maps from $\mathbb{D}$ onto Ω_n normalized by $f_n(0) = w_0$ and $f'(0) > 0$. Then the following conditions hold:*

(1) *If $f_n \to f$ uniformly on compact sets, then $\Omega_n \to \Omega = f(\mathbb{D})$.*
(2) *If $\Omega_n \to \Omega$ in the Carathéodory sense and $\Omega \neq \mathbb{C}$, then $f_n \to f$ uniformly on compact sets, where f is the univalent map onto Ω normalized by $f(0) = w_0$ and $f'(0) > 0$.*
(3) *In both cases described above, if the limit of domains is not a singleton or, equivalently, the limiting map f is not a constant, the inverse functions f_n^{-1} converge to f^{-1} locally uniformly.*

Proof. Let us assume that $f_n \to f$ uniformly on every compact set. By the Hurwitz theorem (Theorem 2.12), f is either constant or univalent.

First, let us consider the case when the limiting function is constant. Since $f_n(0) = w_0$, we have $f(z) = w_0$ for all $z \in \mathbb{D}$. We claim that in this case, the kernel is $\{w_0\}$ and Ω_n converges to the kernel. Let us assume that the kernel is non-trivial, and hence there is $r > 0$ such that $B(w_0, r) \subset \Omega_n$

for almost all n. Hence, by the Schwarz lemma (Theorem 2.2), $|f_n'(0)| \geq r$ for almost all n. Since f_n converges to f uniformly on compact sets, we have $|f'(0)| \geq r > 0$, which contradicts the assumption that f is constant. This proves that the kernel is trivial. Since the same argument holds for every subsequence of f_n, the kernel of every subsequence of (Ω_n) is also trivial. By definition, this implies that the domains converge to the kernel.

Next, we assume that f is a univalent function. We denote the kernel of the sequence (Ω_n) by Ω'. We claim that $\Omega = \Omega'$.

Let w_1 be any point in Ω. Then there is $0 < r < 1$ such that $w_1 \in f(r\mathbb{D})$. It is clear that $f(r\mathbb{D})$ is a domain containing both w_0 and w_1. We claim that this domain is contained in Ω_n for all sufficiently large n. This will imply that $\Omega \subset \Omega'$. The basic idea is very simple: since $f_n \to f$ uniformly, the image of $r\mathbb{D}$ under f_n must be very close to $f(r\mathbb{D})$. The rigorous argument is more involved. Let us consider a circle $\{|z| = R\}$ for some $r < R < 1$. By continuity of f, there is some $\delta > 0$ such that $|f(z) - w'| > \delta$ for every $w' \in f(r\mathbb{D})$ and every z of modulus R. By uniform convergence, there is N such that $|f_n(z) - f(z)| < \delta$ for every $n > N$ and every z on the circle of radius R. Take arbitrary $w' \in f(r\mathbb{D})$ and consider the function $f_n(z) - w' = (f(z) - w') + (f_n(z) - f(z))$. For all $n > N$, the modulus of the first term is larger than the modulus of the second term and hence, by Rouché's theorem, $f - w'$ and $f_n - w'$ have the same number of roots inside the circle of radius R. Since $f = w'$ for exactly one point, the same is true for f_n. This proves that for all $n > N$, we have $f(r\mathbb{D}) \subset f_n(R\mathbb{D}) \subset f_n(\mathbb{D}) = \Omega_n$.

To prove the opposite inclusion, we consider a point w_1 from the kernel. By definition, there is a domain D, which contains both w_0 and w_1, and is inside Ω_n for all n larger than some N. The inverse function $g_n(w) = f_n^{-1}(w)$ is defined on D for $n > N$ and $|g_n(w)| < 1$. By the Montel theorem (Theorem 2.10), functions g_n on $\mathbb{D}$ form a normal family and we can choose a subsequence g_{n_k} which converges locally uniformly to some function g with $|g| < 1$. In particular, g_{n_k} converges to g uniformly in some neighbourhood of w_1. By definition, $w_1 = f_{n_k}(g_{n_k}(w_1))$ and, passing to the limit, we get $w_1 = f(g(w_1)) \in f(\mathbb{D})$. This completes the proof that $\Omega = \Omega'$.

As before, the same argument is valid for every subsequence of f_n; hence $f(\mathbb{D})$ is not only the kernel of (Ω_n), but it is also the kernel of every subsequence, that is, $\Omega_n \to f(\mathbb{D})$ in the sense of Carathéodory. This completes the proof of the first part. The argument about the convergence of g_n gives the third part of the theorem.

To prove the second part, we assume that Ω_n converges to the kernel $\Omega \neq \mathbb{C}$. By the Koebe distortion theorem (Theorem 3.7), the disc

$B(w_0, |f_n'(0)|/4)$ is inside Ω_n. In particular, if derivatives at zero are unbounded, then there is a subsequence of (Ω_n) with the kernel equal to $\mathbb{C}$, which contradicts our assumptions. Hence, the derivatives at zero are uniformly bounded. Combined with the Growth theorem (Theorem 3.11), this implies that the functions f_n are uniformly bounded on every compact set, and, again by the Montel theorem, form a normal family.

From a normal family, we can always choose a convergent subsequence and the standard argument tells us that either the sequence is convergent or there are two subsequences with different limits. Now, let us assume that there are two subsequences f_{n_k} and f_{m_k} that converge to f and $\tilde{f}$, respectively. The first part of the proof can be applied to show that $f(\mathbb{D})$ and $\tilde{f}(\mathbb{D})$ are kernels of (Ω_{n_k}) and (Ω_{m_k}). By definition of kernel convergence, the kernels of subsequences must coincide and $f(\mathbb{D}) = \tilde{f}(\mathbb{D})$. Since both f and $\tilde{f}$ map zero to w_0 and $f'(0) \geq 0$, $\tilde{f}'(0) \geq 0$, the uniqueness part of the Riemann mapping theorem implies that $f = \tilde{f}$, which contradicts our assumption, and proves that the sequence f_n converges to some function f. By the first part of the proof, $\Omega = f(\mathbb{D})$. □

With the notion of Carathéodory convergence, we can now state a rather simple but important result that we will need later. Since we now have a suitable notion of convergence and topology on the space of domains, we can ask the question: Is it true that every domain can be approximated by 'nice' domains? Obviously, the answer depends on what we mean by 'nice'.

First, we note that every domain can be approximated by domains such that their boundaries are analytic curves. Using linear transformation, we can reduce the problem to the question about the function from the class S. Let $f \in S$ and $\Omega = f(\mathbb{D})$ be the corresponding domain. Let us consider a map $f_r(z) = f(rz)/r$ for $r \in (0, 1)$. By considering the power series, it is clear that $f_r \in S$. The boundary of $\Omega_r = f_r(\mathbb{D})$ is the rescaled image of the circle of radius r under the map f. Since $r < 1$, this image is an analytic curve. There are several ways to check that f_r converges to f uniformly on compacts as $r \to 1$. The simplest way is to note that the power series converges uniformly inside the radius of convergence. This means that if we assume that $|z| \leq \rho < 1$, then the series

$$f(z) - f_r(z) = \sum_{n=2}^{\infty} a_n z^n (1 - r^{n-1})$$

converges uniformly, so we can pass to the limit term by term and show that $f - f_r \to 0$ uniformly on $|z| < \rho$.

This approximation by analytic domains is already a nice property, but later on we will need a slightly different class of *slit domains.*

Definition 3.17. A *slit domain* is a domain of the form $\mathbb{C} \setminus \bigcup J_i$ where J_i are disjoint Jordan curves. A *single-slit domain* is the complement of a single Jordan curve. Since we require the domain to be simply connected, this curve must extend to infinity. A *single-slit mapping* is a conformal map onto a single-slit domain.

We are mostly interested in the single-slit mappings from S. The following theorem proves that single-slit maps are dense in S in the sense of Carathéodory convergence.

Theorem 3.18. *Let f be a univalent map from S. Then there is a sequence of single-slit maps $f_n \in S$ such that $f_n \to f$ uniformly on compacts.*

Proof. Let us fix some $\rho < 1$. From the argument preceding the theorem, we know that for every $\epsilon > 0$, there is $r = r(\epsilon) < 1$ such that $f_r(z) = f(rz)/r$ is uniformly ϵ-close to f for $|z| < \rho$. Let us fix this r and denote the corresponding function f_r by $\tilde{f}$.

The image $\tilde{f}(\mathbb{D})$ is a domain bounded by an analytic Jordan curve γ. The curve γ can be naturally parametrized by $\gamma(\theta) = \tilde{f}(e^{i\theta})$ with $\theta \in [0, 2\pi]$. We define a curve γ_δ as the restriction of γ, namely, we restrict θ to the interval $[\delta, 2\pi]$. This curve connects $w_\delta = \gamma(\delta)$ with $\gamma(2\pi) = w_0$. Since γ is a simple closed Jordan curve, there is a Jordan curve $\tilde{\gamma}$ which lies completely outside of γ and connects $w_0 = \gamma(0) = \gamma(2\pi)$ with infinity. Now, consider a domain $\Omega_\delta = \mathbb{C} \setminus (\gamma_\delta \cup \tilde{\gamma})$: see Figure 3.3. By construction, Ω_δ is a single-slit domain. Let h_δ be the unique univalent map $h_\delta : \mathbb{D} \to \Omega_\delta$ such that $h_\delta(0) = 0$ and $h'_\delta(0) > 0$.

It is easy to see that the kernel of $(\Omega_\delta)_{\delta>0}$ with respect to the origin is $\tilde{f}(\mathbb{D})$. By the Carathéodory convergence theorem, functions $h_\delta \to \tilde{f}$ uniformly on all compacts. From the uniform convergence, we have that $h'_\delta(0) \to \tilde{f}'(0) = 1$; hence the functions $\tilde{h}_\delta(z) = h_\delta(z)/h'_\delta(0)$, which belong to the class S, also converge to $\tilde{f}$ uniformly on compacts. In particular, we can choose a sufficiently small δ such that $|\tilde{h}_\delta - \tilde{f}| < \epsilon$ on $|z| < \rho$.

This proves that for every $0 < \rho < 1$ and every $\epsilon > 0$, there is a single-slit map from S which is ϵ-close to f on $|z| < \rho$. Choosing sequences $\epsilon_n \to 0$ and $\rho_n \to 1$, there are single-slit maps f_n such that $|f(z) - f_n(z)| < \epsilon_n$ on $|z| < \rho_n$. This is equivalent to the statement of the theorem. □

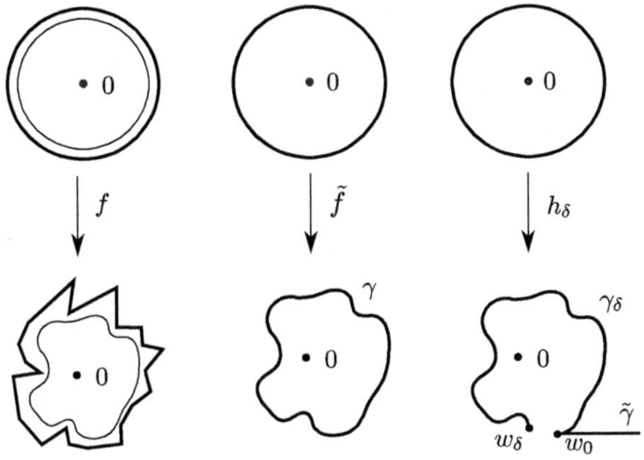

Fig. 3.3. Any map from S can be approximated by a single-slit map.

Essentially, the same proof gives that every domain $\Omega \subset \mathbb{D}$ can be approximated by single-slit disc domains, that is, domains of the form $\mathbb{D}\setminus\gamma$, where γ is a Jordan curve started on the unit circle.

Both the notion of the Carathéodory convergence and the corresponding convergence theorem can be generalized to many other settings, in particular to non-simply connected domains and to maps with different normalization. It is not our aim to present a comprehensive study, but there is one setup that will be useful later on.

Let (K_n) be a sequence of half-plane hulls. We work with sequences for simplicity, but essentially the same argument will work for K_t, $t \in \mathbb{R}$. Let us assume that they are uniformly bounded, that is, there is $R > 0$ such that all K_n are inside $R\mathbb{D}$. By f_n, we denote the corresponding functions f_{K_n} that have the thermodynamic normalization and map $\mathbb{H}$ onto $\Omega_n = \mathbb{H}\setminus K_n$. It is convenient to identify f_n with their symmetric extensions that map $\widehat{\mathbb{C}}\setminus I_n$ onto $\widehat{\mathbb{C}}\setminus(K_n \cup \bar{K}_n)$, where I_n is a collection of real intervals. Adding a part of the real line to K_n does not change the maps at all, but it allows us to assume that all I_n are single intervals. Moreover, since all K_n are bounded, the corresponding intervals $I_n \in [-2R, 2R]$. By extending K_n further along the real line, we can assume that all I_n are equal to $[-2R, 2R]$. This allows us to assume that all functions f_n are defined on the same domain $\widehat{\mathbb{C}}\setminus[-2R, 2R]$.

In this setup, we can define the kernel of (Ω_n) (whether it is symmetrized or not) in exactly the same way as before, as the largest unbounded simply connected domain such that its every closed subset is contained in all Ω_n

for sufficiently large n. We say that $\Omega_n \to \Omega$ (in the Carathéodory sense) if Ω is the kernel of every subsequence. We say that hull $K_n \to K$ if the corresponding domains converge.

Note that our boundedness assumption excludes the possibility of the trivial kernel. But it is still possible that the limiting domain is $\mathbb{H}$ and $K = \emptyset$ (in the symmetrized and extended case, the limiting domain is $\widehat{\mathbb{C}} \setminus [-2R, R]$ and $K = [-2R, R]$). For example, this happens when $K_n = \{z \in \mathbb{H} : |z| \leq 1/n\}$. Also note that in this case the maps f_n do not blow up as in the case of the usual kernel being $\mathbb{C}$, but converge to the identity. This happens due to the thermodynamic normalization at infinity. As before, the Carathéodory limit might be different from the topological one.

Exercise 3.12. Let K_t be the arc $\gamma([0,t])$, where $\gamma(t) = \exp(it)$ as in Exercise 3.18. The hull K_t is a simple curve for $t < \pi$. Show that the kernel of $\Omega_t = \mathbb{H} \setminus K_t$ is $\mathbb{H} \setminus \mathbb{D}$ and the half-disc is the limit of K_t. By considering the maps $g_t = g_{K_t}$ show directly that they converge to $g(z) = z + 1/z$, which is the mapping out of the half-disc.

In the half-plane case, the Carathéodory convergence theorem has the following form.

Theorem 3.19. *Let $\Omega_n = \mathbb{H} \setminus K_n$ and f_n be as above. Domains Ω_n converge to $\Omega = \mathbb{H} \setminus K$ if and only if $f_n \to f = f_K$ uniformly away from the real line. In the case of convergence, the inverse maps g_n converge uniformly on closed subsets of Ω.*

The theorem can be proved by following exactly the same argument as in the disc setting, the main difference being that one has to either use the spherical version of normal families or use normality of $f_n(z) - z$. Alternatively, we can consider functions $\tilde{f}_n = 1/f_n(RJ)$, where J is the Joukowsky map (here, we use the symmetrized version of f_n). These functions map $\mathbb{D}$ onto simply connected domains $\widetilde{\Omega}_n$. It is easy to show that the convergence of Ω_n is equivalent to the convergence of $\widetilde{\Omega}_n$ and the local uniform convergence of f_n is equivalent to the local uniform convergence of $\tilde{f}_n$. The details are left to the reader.

Remark 3.20. Following the same lines, one can show as in the disc case that any half-plane hull can be approximated in the Carathéodory sense by single-slit hulls, that is, by hulls that are simple Jordan curves. Another useful observation is that the family of maps $f_K(z) - z$, where

K is a hull inside $R\mathbb{D}$, is normal, and hence from every sequence we can choose a convergent sequence. This implies that there is a sequence f_n converging locally uniformly to $f = f_K$ (by normalization at infinity, the map cannot be constant). By the Carathéodory theorem, this implies that $K_n \to K$ in the Carathéodory sense. This proves that the space of uniformly bounded hulls is compact with respect to the Carathéodory convergence.

Remark 3.21. If domains $\Omega_n = \mathbb{H}\backslash K_n$ are decreasing and the corresponding hulls are increasing and uniformly bounded, then the Carathéodory limit of Ω_n is the unbounded component of $\mathbb{H} \setminus \bigcup K_n$. It is natural to assume that the limit of K_n is the 'hole filling' of the topological limit $\bigcup K_n$, namely, we add to it all bounded components of the complement. A typical example is given in Exercise 3.12; see also Figure 3.5.

3.4. Subordination and the Carathéodory Class

We have already seen in Lemma 2.4 an example of an inequality connecting univalent maps of a domain and its subdomain. In this section, we extend these results; in particular, in some instances, we will be able to drop the univalence assumption and derive several useful implications. As in Lemma 2.4, these results are closely related to the Schwarz lemma.

Definition 3.22. Let functions f and g be analytic in the unit disc $\mathbb{D}$. We say that f is *subordinate* to g (denoted $f \prec g$) if there is an analytic function ϕ in the unit disc such that $\phi(0) = 0$, $|\phi(z)| < 1$, and

$$f(z) = g(\phi(z)), \quad z \in \mathbb{D}.$$

It follows immediately from the definition that if $f \prec g$, then $f(0) = g(0)$ and $f(\mathbb{D}) \subset g(\mathbb{D})$. Note that in this definition, unlike in most of this book, we do not assume that functions f, g, and ϕ are univalent. If the function g is univalent, then it is easy to check that $f \prec g$ if and only if $f(0) = g(0)$ and $f(\mathbb{D}) \subset g(\mathbb{D})$. Indeed, in this case, we can see that $f \prec g$ by considering the function $\phi(z) = g^{-1}(f(z))$.

In the case when both f and g are univalent, the subordination is particularly simple, since ϕ is also univalent. This is exactly the setup of Lemma 2.4 which gives an inequality on f^{-1} and g^{-1}.

Without the univalence assumption, the picture is more complicated, but it is still possible to derive some estimates which could be interpreted as generalizations of the Schwarz lemma.

Lemma 3.23. *Let f be subordinate to g. Then*

$$f(r\mathbb{D}) \subset g(r\mathbb{D}) \quad \textit{for all } 0 \le r \le 1 \tag{3.11}$$

which immediately implies

$$\max_{|z|\le r} |f(z)| \le \max_{|z|\le r} |g(z)|. \tag{3.12}$$

Moreover,

$$\max_{|z|\le r}(1-|z|^2)|f'(z)| \le \max_{|z|\le r}(1-|z|^2)|g'(z)|. \tag{3.13}$$

Proof. The first part is a direct corollary of Lemma 2.2. Indeed, it implies that $|\phi(z)| \le |z|$, which implies $\phi(r\mathbb{D}) \subset r\mathbb{D}$; since $f = g(\phi)$, we immediately have $f(r\mathbb{D}) = g(\phi(r\mathbb{D})) \subset g(r\mathbb{D})$.

To prove the second inequality, we apply the chain rule

$$(1-|z|^2)|f'(z)| = (1-|z|^2)|g'(\phi(z))||\phi'(z)| \le |g'(\phi(z))|(1-|\phi(z)|^2),$$

where the last inequality is valid by the second part of Theorem 2.3, which states that $|\phi'(z)| \le (1-|\phi(z)|^2)/(1-|z|^2)$. Taking the supremum over $|z| \le r$, we have

$$\begin{aligned}\sup_{|z|\le r}(1-|z|^2)|f'(z)|| &\le \sup_{|z|\le r} |g'(\phi(z))|(1-|\phi(z)|^2)\\ &= \sup_{w\in\phi(r\mathbb{D})} |g'(w)|(1-|w|^2)\\ &\le \sup_{|w|\le r} |g'(w)|(1-|w|^2),\end{aligned}$$

where the last inequality holds since $\phi(r\mathbb{D}) \subset r\mathbb{D}$. □

There is a particularly important class of functions that could be defined in terms of subordination — Carathéodory class $\mathcal{P}$. These functions will appear naturally in the theory of Loewner evolution, which we will develop in Chapter 5. It is usually defined as the class of all functions $p(z)$ that are analytic in the unit disc $\mathbb{D}$ satisfying two conditions:

$$p(0) = 1 \quad \text{and} \quad \operatorname{Re} p(z) > 0. \tag{3.14}$$

From the above discussion, it is easy to see that these conditions are equivalent to

$$p(z) \prec g(z) = \frac{1+z}{1-z},$$

where g is a Möbius transformation from $\mathbb{D}$ onto the right half-plane $\{z : \operatorname{Re}(z) > 0\}$.

In this particular case, the general subordination estimates (3.11) and (3.13) give us

$$\frac{1-|z|}{1+|z|} \le |p(z)| \le \frac{1+|z|}{1-|z|} \tag{3.15}$$

and

$$|p'(z)| \le \frac{2}{(1-|z|)^2}. \tag{3.16}$$

Exercise 3.13. Show that the Carathéodory class $\mathcal{P}$ is compact.

One of the nice and important properties of functions from the class $\mathcal{P}$ is that they have an integral representation very similar to the Schwarz formula (2.6).

Theorem 3.24 (Herglotz). *Let $p(z) = 1+a_1z+a_2z^2+\cdots$ be an analytic function in the unit disc $\mathbb{D}$. Then $p \in \mathcal{P}$ if and only if there is a non-decreasing function $h(\theta)$ on $[0,2\pi]$ such that $h(2\pi) - h(0) = 1$ and*

$$p(z) = \int_0^{2\pi} \frac{e^{i\theta}+z}{e^{i\theta}-z} \mathrm{d}h(\theta). \tag{3.17}$$

In the classical complex analysis literature, the integral representation (3.17) is usually given in terms of the Stieltjes integral. Alternatively, the integral can be rewritten in terms of a probability measure μ on $[0,2\pi]$ or on the unit circle such that h is its cumulative distribution function. In this case, (3.17) becomes

$$p(z) = \int_0^{2\pi} \frac{e^{i\theta}+z}{e^{i\theta}-z} \mu(\mathrm{d}\theta). \tag{3.18}$$

Proof of Theorem 3.24. Let us first assume that $p \in \mathcal{P}$ and fix some $0 < r < 1$. For this value of r, we can apply the Schwarz formula (2.7) to the rescaled function $p_r(z) = p(rz)$ which is continuous in the closed unit disc:

$$\begin{aligned} p_r(z) &= \frac{1}{2\pi} \int_0^{2\pi} \frac{e^{i\theta}+z}{e^{i\theta}-z} \operatorname{Re}(p_r(e^{i\theta}))\mathrm{d}\theta \\ &= \int_0^{2\pi} \frac{e^{i\theta}+z}{e^{i\theta}-z} \mathrm{d}h_r(\theta) = \int_0^{2\pi} \frac{e^{i\theta}+z}{e^{i\theta}-z} \mu_r(\mathrm{d}\theta), \end{aligned}$$

where

$$h_r(\theta) = \frac{1}{2\pi} \int_0^{\theta} \operatorname{Re}(p_r(e^{it}))\mathrm{d}t$$

and μ_r is the corresponding probability measure.

Let r_n be any sequence of radii such that $0 \leq r_n < 1$ and $r_n \to 1$. Helly's selection theorem (see, for example, Natanson (2016, Lemma 2 in VIII.4)) states that from every sequence of uniformly bounded monotone functions, one can find a subsequence converging pointwise. Hence, there is a subsequence r_{n_k} and a non-decreasing function $h(\theta)$ such that for this subsequence, $h_r(\theta) \to h(\theta)$ at all points where h is continuous. Passing to the limit in the integral representation for p_r we get (3.17).

Alternatively, we can use that the space of probability measures on $[0, 2\pi]$ is compact in weak topology (Prokhorov's theorem, see, for example, Koralov and Sinai (2007, Lemma 8.10)) and choose a convergent subsequence of μ_{r_n}. Passing to the limit in the above integral representation, we obtain (3.18).

The other direction of the theorem is almost trivial. Indeed, the integrand in (3.17) or (3.18) has a positive real part, hence $\mathrm{Re}\, p > 0$. The condition $p(0) = 1$ follows immediately from $h(2\pi) - h(0) = 1$ or, equivalently, from the fact that μ is a probability measure. □

3.5. Capacities and Geometry

In the previous parts, we have introduced several notions that have a clear geometric flavour: conformal radius, logarithmic capacity, and half-plane capacity. In this section, we study their properties.

Conformal radius and logarithmic capacity. The first two notions are given in terms of the derivative of Riemann maps and thus are very closely related. Indeed, let Ω be a simply connected domain, z_0 be a point in Ω, and $f : \mathbb{D} \to \Omega$ be a conformal map with $f(0) = z_0$. By the definition of the conformal radius at z_0, $\rho(z_0, \Omega) = |f'(0)|$. Now, let us apply $1/(z - z_0)$ to Ω and denote the image by $\widetilde{\Omega}$ and set K to be $\mathbb{C} \setminus \widetilde{\Omega}$. The map $\tilde{f}(z) = 1/(f(1/z) - z_0)$ is a conformal map from $\mathbb{D}_-$ onto $\widetilde{\Omega}$ and a simple computation gives that $\tilde{f}'(\infty) = 1/|f'(0)|$ so that $\mathrm{cap}(K) = -\log(\rho(z_0, \Omega))$.

We also notice that the map $1/(z - z_0)$ is very simple, so any geometric statement about Ω can be translated into an equivalent statement about K. This shows that there is a direct and simple way to translate any result connecting geometry and conformal radius into an equivalent statement about geometry and logarithmic capacity.

In the framework of conformal mappings, the logarithmic capacity provides a convenient method to estimate the size of a compact set K. Indeed, since it is formulated in terms of conformal maps, one can use a wealth of analytic tools to compute or estimate it; on the other hand, Corollary 3.6 immediately implies that the exponent of capacity is proportional to the diameter of K. For the sake of completeness, we give the precise statement.

Proposition 3.25. *Let K be a compact set containing more than one point such that $\Omega = \widehat{\mathbb{C}} \setminus K$ is simply connected. Then*

$$1 \leq \frac{\mathrm{rad}(K)}{\exp(\mathrm{cap}(K))} \leq 4,$$

where $\mathrm{rad}(K)$ *is the radius of the smallest disc containing K.*

The bounds in the proposition are sharp. The lower bound is obtained when K is a disc, the upper bound is obtained when K is a straight interval.

Half-plane capacity. In the half-plane case, the relation between capacity and the Euclidean size of K is not that straightforward. We start by considering a couple of simple examples.

Exercise 3.14. Let K be a half disc $\{z : \mathrm{Im}\, z \geq 0,\ |z| \leq R\}$. Show that $\mathrm{hcap}(K) = R^2$.

Exercise 3.15. Let K be a vertical interval $[0, it]$. Show that $\mathrm{hcap}(K) = t^2/2$.

From these two examples, we can see that the half-plane capacity has square scaling. It is easy to see that this is true for general hulls as well.

Proposition 3.26. *Let K be a half-plane hull and let $\lambda K = \{z : z/\lambda \in K\}$ be its rescaling by a positive factor λ. Then $g_{\lambda K}(z) = \lambda g_K(z/\lambda)$ and* $\mathrm{hcap}(\lambda K) = \lambda^2 \,\mathrm{hcap}(K)$

Proof. The proof is absolutely straightforward. It is easy to see that $\lambda g_K(z/\lambda)$ indeed maps the complement of λK onto $\mathbb{H}$. Expanding it at infinity, we have

$$\lambda g_K(z/\lambda) = z + \frac{\lambda^2\,\mathrm{hcap}(K)}{z} + \cdots,$$

which immediately implies that this map is indeed the same as $g_{\lambda K}(z)$ and that $\mathrm{hcap}(\lambda K) = \lambda^2\,\mathrm{hcap}(K)$.

Exercise 3.16. Prove the following properties of the half-plane capacity:

(1) Half-plane capacity is translationally invariant, namely $\operatorname{hcap}(K) = \operatorname{hcap}(K+x)$ for any real x.
(2) Let $K_1 \subset K_2$ be two half-plane hulls. Then

$$\operatorname{hcap}(K_2) = \operatorname{hcap}(K_1) + \operatorname{hcap}(\tilde{K}), \tag{3.19}$$

where $\tilde{K} = g_{K_1}(K_2 \setminus K_1)$.

Proposition 3.27. *The half-plane capacity is continuous with respect to the Carathéodory convergence, namely, let K_n be uniformly bounded hulls that converge in the sense of Carathéodory to K. Then* $\operatorname{hcap}(K_n) \to \operatorname{hcap}(K)$.

Proof. Let $g_n : \mathbb{H} \setminus K_n \to \mathbb{H}$ and $g : \mathbb{H} \setminus K \to \mathbb{H}$ be the corresponding thermodynamically normalized maps. By the Carathéodory convergence theorem, the corresponding maps $g_n \to g$ uniformly away from the real line, in particular, near infinity. This implies that the coefficients in the power series also converge. Since the half-plane capacity is defined in terms of coefficients, it also converges. □

By definition, half-plane capacity $\operatorname{hcap}(K)$ is given in terms of behaviour of g_K at infinity. But from the Schwarz formula, we can express $f_K = g_K^{-1}$ in the upper half-plane in terms of its boundary values. This gives us a very useful integral representation of capacity.

Lemma 3.28. *Let K be a half-plane hull, $\Omega = \mathbb{H} \setminus K$ and $f_K : \mathbb{H} \to \Omega$ be the corresponding conformal map with thermodynamic normalization. Then f_K has expansion at infinity*

$$f_K(z) = z - \frac{\operatorname{hcap}(K)}{z} + \cdots$$

and

$$\operatorname{hcap}(K) = \frac{1}{\pi} \int_{-\infty}^{\infty} \operatorname{Im} f_K(x) \mathrm{d}x. \tag{3.20}$$

Proof. The first formula follows immediately from the thermodynamic normalization of g_K and the fact that f_K is inverse of g_K.

By our choice of normalization, we can apply the Schwarz formula[1] (2.10) to $f_K(z) - z$ to get

$$f_K(z) - z = \frac{1}{\pi}\int_{-\infty}^{\infty} \frac{\operatorname{Im}(f_K(x) - x)}{x - z}\mathrm{d}x = \frac{1}{\pi}\int_{-\infty}^{\infty} \frac{\operatorname{Im} f_K(x)}{x - z}\mathrm{d}x. \quad (3.21)$$

Next, we multiply by z and pass to the limit as $z \to \infty$. The left-hand side converges to $-\operatorname{hcap}(K)$, and the right-hand side can be written as

$$\lim_{z\to\infty} \frac{1}{\pi}\int_{-\infty}^{\infty} \left(-1 + \frac{x}{x - z}\right) \operatorname{Im} f_K(x)\mathrm{d}x.$$

Note that since K is compact, $\operatorname{Im} f_K(x) = 0$ outside of a compact set. This allows to pass to the limit under the integral sign and prove (3.20). □

By construction, $\operatorname{Im} f_K(x) \geq 0$ and, unless $f_K(z) = z$, there are points where it is strictly positive. This immediately gives the following corollary.

Corollary 3.29. *Let K be a non-trivial hull in $\mathbb{H}$, that is, $\Omega_K = \mathbb{H} \setminus K \neq \mathbb{H}$. Then* $\operatorname{hcap}(K) > 0$. *Let K_1 and K_2 be two hulls such that $\Omega_{K_2} \subsetneq \Omega_{K_1}$. Then* $\operatorname{hcap}(K_2) > \operatorname{hcap}(K_1)$.

Proof. We have already explained the first part. The second part follows from the first one and additivity of capacity (3.19). □

In some sense, the half-plane capacity measures the size of the compact set K as 'seen from infinity'. The relations between the capacity and geometrical notions of size are not so simple. It is easy to show that an upper bound on the size implies an upper bound on capacity. To be more precise, let us define $\operatorname{rad}(K)$ to be the radius of the smallest half-disc containing K:

$$\inf\{r > 0 : \text{ there is } x \in \mathbb{R} \text{ such that } K \subset B(x, r)\}.$$

By monotonicity of capacity, the capacity of K is bounded by the capacity of the half-disc of radius $\operatorname{rad}(K)$ which is equal to $\operatorname{rad}(K)^2$.

The next exercise shows that there is no lower bound on the capacity in terms of $\operatorname{rad}(K)$.

Exercise 3.17. Suppose that $0 < \alpha < 1$ and consider the function

$$f(z) = (z + \alpha)^{1-\alpha}(z + \alpha - 1)^{\alpha},$$

[1]We proved the formula for functions that are continuous up to the boundary, f_K is not necessarily continuous, but it has vertical limits almost everywhere, which is sufficient for the integral formula to hold.

where we choose the branch which is positive for $z > 1-\alpha$. Show that $f = f_K$, where K is the interval $[0, \alpha^\alpha(1-\alpha)^{1-\alpha}e^{i\alpha\pi}]$. Show that it maps $z = 1-2\alpha$ to the endpoint of K. Use this to show that

$$\mathrm{hcap}(I_\alpha) = \frac{1}{2}\alpha^{1-2\alpha}(1-\alpha)^{2\alpha-1},$$

where I_α is the interval $[0, e^{i\alpha\pi}]$.

Note that $\mathrm{hcap}(I_\alpha) \to 0$ as $\alpha \to 0$ or $\alpha \to 1$. This proves that there is no lower bound on the capacity of K in terms of $\mathrm{rad}(K)$.

Exercise 3.18. For $0 \le \theta \le \pi$ let $\Omega = \Omega_\theta = \mathbb{H}\setminus\gamma([0,\theta])$, where $\gamma(t) = e^{it}$. Find the thermodynamically normalized map $g_\theta : \Omega \to \mathbb{H}$ and the half-plane capacity of $K_\theta = \gamma([0,\theta])$.

3.6. Further Properties of Mapping-Out Functions

In this section, we study the properties of mapping-out functions g_K. This class of functions is closely related to the class Σ but there are some important distinctions.

We start with similarities. As we discussed before, the function g_K can be extended by the Schwarz reflection to a function from $\widehat{\mathbb{C}} \setminus (K \cup \bar{K})$ onto $\widehat{\mathbb{C}} \setminus I_K$, where $\bar{K}$ is the complex conjugate of K and I_K is a collection of real intervals. Identifying g_K with this extension, we see that all functions g_K belong to the class of univalent maps that preserve infinity and are normalized to have the expansion $z + O(z^{-1})$ at infinity. In this sense, these functions are similar to functions in class Σ. But there are important differences. First, for different K, functions g_K as well as f_K are defined in different domains. This is not such a big problem since many results of functions in S and Σ could be reformulated for general domains. A good example is Theorem 3.7. Another important difference is that we do not assume that K is connected. This implies that the (extended) map g_K might be a conformal map between multiply connected domains.

Our first result is a simple corollary of (3.21) and is in the same spirit as the Schwarz lemma and Lemma 2.4.

Lemma 3.30. *Let K be a half-plane hull and g_K be a corresponding mapping. Suppose that $x \in \mathbb{R}$ is to the right of K, that is, $[x, \infty)$ does not intersect K. Then*

$$g_K(x) > x. \tag{3.22}$$

For points that are to the left of K, the inequality is reversed.

This lemma has a simple geometric interpretation: the mapping-out function pushes the hull down and it pushes the real line away towards infinity. The proof is a straightforward application of the Schwarz integral formula applied to f_K and is left as an exercise.

Exercise 3.19. Use formula (3.21) to prove Lemma 3.30.

Lemma 3.30 can be used to obtain two similar results: one states that $g_K(x)$ is monotone in K and the other gives an upper bound on $g_K(x) - x$, given a bound on K.

Lemma 3.31. *Suppose that $K_1 \subset K_2$ are two hulls and let x be a point to the right of K_2. Then*

$$g_{K_1}(x) \leq g_{K_2}(x), \tag{3.23}$$

where equality happens for some x if and only if $K_1 = K_2$.

Proof. Let us define K to be the closure of $g_{K_1}(K_2 \setminus K_1)$, which is a compact hull unless $K_1 = K_2$. Assuming that this is the case, we consider the function g_K. By uniqueness of the Riemann map with thermodynamic normalization, $g_{K_2}(z) = g_K(g_{K_1}(z))$. Applying (3.22) to g_K, we get

$$g_{K_2}(x) > g_{K_1}(x).$$

□

Corollary 3.32. *Let us assume that a non-trivial compact hull K is inside the unit disc $\mathbb{D}$ and that $x > 1$. Then*

$$x < g_K(x) < x + \frac{1}{x}. \tag{3.24}$$

For $x < -1$, we have $x > g_K(x) > x + 1/x$.

Proof. This result is a combination of the previous lemma applied to $K_1 = K$ and $K_2 = \mathbb{D} \cap \mathbb{H}$ and a simple observation that $g_{\mathbb{D}}(z) = z + 1/z$. □

Next, we show that a similar result holds in the upper half-plane as well, namely, there is a *uniform* estimate for $|g_K(z) - z|$ given a bound on K.

Lemma 3.33. *Let K be a half-plane hull, $\Omega = \mathbb{H} \setminus K$ and $g_K : \Omega \to \mathbb{H}$ be the corresponding conformal map. Then*

$$|g_K(z) - z| \leq 3\,\mathrm{rad}(K), \tag{3.25}$$

for all $z \in \Omega$.

Proof. We start by assuming that $K \subset \mathbb{D}$ and consider $(x^+,\infty) = g_K((1,\infty))$. By Corollary 3.32, $x^+ \in [1,2]$. For all $x > 1$, we have $0 \leq g_K(x) - x \leq 1/x \leq 1$, or, equivalently, $|f_K(x) - x| \leq 1$ for all $x \geq x^+$. In a similar way, we define x^- by $(-\infty, x^-) = g_K((-\infty,-1))$ and by the same argument $|f_K(x) - x| \leq 1$ for $x < x^-$. We already know that $(x^-, x^+) \subset (-2,2)$. Since $K \subset \mathbb{D}$, the image $f_K((x^-,x^+))$ is inside $\mathbb{D}$; this proves that $|f_K(x)-x| \leq 3$ on (x^-,x^+). Combining all these estimates, we can see that $|f_K(x) - x| \leq 3$ and since $f_K(z) - z \to 0$ at infinity, by the maximum modulus principle $|f_K(z) - z| \leq 3$ everywhere in $\mathbb{H}$. This is equivalent to $|g_K(z) - z| \leq 3$ in the complement of K.

Finally, to prove the general case, we assume that

$$K \subset \{z \in \mathbb{H} : |z - \zeta| \leq r\}$$

for some $r > 0$ and $\zeta \in \mathbb{R}$. Let $\tilde{K} = (K - \zeta)/r$ be the rescaled and shifted hull. By basic properties of mapping-out functions,

$$g_K(z) = r g_{\tilde{K}}\left((z - \zeta)/r\right) + \zeta.$$

Combining this formula with an estimate $|g_{\tilde{K}}(z) - z| \leq 3$, we prove (3.25). □

With a bit more work, we can get the uniform bound for the next term as well.

Lemma 3.34. *Let K be a half-plane hull which is inside $|z| \leq R$ and $z \in \mathbb{H}$ with $|z| \geq 10R$. Then*

$$\left| g_K(z) - z - \frac{\mathrm{hcap}(K)}{z} \right| \leq \frac{10R\,\mathrm{hcap}(K)}{|z|^2}. \tag{3.26}$$

Constant 10 appearing in this lemma is not sharp; it could be improved, but the particular value is not really important. Also note that the condition that K is inside the disc of radius R is very similar to $\mathrm{rad}(K) \leq R$, but we assume that the centre of the disc is at the origin. Without this assumption, there is no universal estimate in terms of $\mathrm{rad}(K)$. Indeed, let us consider K to be the half-disc of radius 1 centred at x_0. Then $g_K(z) = z + (z - x_0)^{-1}$ and

$$|g(z) - z - 1/z| = \frac{|x_0|}{|z - x_0||z|},$$

which is not uniformly bounded by $c/|z|$ for any c.

Proof of Lemma 3.34. Formulas (3.21) and (3.20) give

$$g_K(z) - z - \frac{\mathrm{hcap}(K)}{z} = \frac{1}{\pi}\int_{-\infty}^{\infty} \mathrm{Im}\, f_K(x)\left(\frac{1}{g_K(z)-x} - \frac{1}{z}\right)\mathrm{d}x.$$

Since $|z| > 10R$ and $|g_K(z) - z| \le 3R$ by (3.25), for $|x| \le 2R$, we have

$$\left|\frac{1}{g_K(z)-x} - \frac{1}{z}\right| = \frac{|g_K(z) - x - z|}{|z||g_K(z) - x|} \le \frac{5R}{|z||z - 5R|} \le \frac{10R}{|z|^2}.$$

By comparing K with the half-disc of radius R, we see that $\mathrm{Im}\, f_K(x) = 0$ if $|x| > 2R$. Combining this with previous estimates, we obtain

$$\left|g_K(z) - z - \frac{\mathrm{hcap}(K)}{z}\right| \le \frac{10R}{|z|^2}\frac{1}{\pi}\int_{-\infty}^{\infty} \mathrm{Im}\, f_K(x)\mathrm{d}x = \frac{10R\,\mathrm{hcap}(K)}{|z|^2}. \quad \square$$

Finally, we would like to mention that these rather sharp estimates of conformal maps can be used to show that the half-plane capacity is comparable to a more geometric quantity. To explain it, we first have to define the Whitney decomposition of $\mathbb{H}$. This is the decomposition of $\mathbb{H}$ into the union of dyadic squares of the form $Q_{k,n} = 2^n((k+i) + [0,1] \times [0,1])$ with $k, n \in \mathbb{Z}$. These are exactly the dyadic squares such that their distance to the real line is equal to their size. The half-plane capacity of a hull K is comparable to the Euclidean area of the union of all Whitney squares that intersect K; see Figure 3.4. We are not going to prove this result; interested readers can find the proof in Rohde and Wong (2014) or Lalley *et al.* (2009).

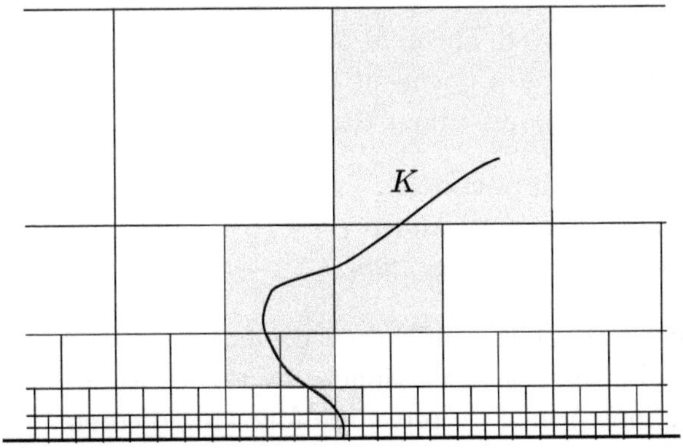

Fig. 3.4. The shaded area is the union of all Whitney squares that are intersected by K. Their total area is comparable to hcap(K).

3.7. Coefficient Problems

Taylor coefficients of an analytic function give complete information about the function, but it is usually very difficult to extract this information. In particular, if we know that the function is univalent, then it is not clear what can be said about its coefficients. The other direction is even more difficult: given coefficients, it is extremely difficult even to determine whether the corresponding function is univalent or not.

Some results in this direction are rather simple. We have already seen that Gronwall's theorem gives an upper bound on the coefficients b_n for functions from class Σ, namely $|b_n| \leq n^{-1/2}$. This estimate is sharp for $n = 1$, but not sharp for $n \geq 2$. Together with the square root transform, this implies the sharp estimate on the second coefficient for functions from class S.

Pursuit of universal coefficient inequalities has motivated a lot of research in geometric function theory and these inequalities were used to prove many interesting results that are not directly related to coefficients. The above-discussed Growth, Rotation, and Distortion theorems are very good examples to show that this approach is very fruitful.

In this section, we discuss (mostly without proofs and paying more attention to the history) the coefficient inequalities and related questions for functions from classes S and Σ. Extensive discussion of coefficients and related problems can be found in Duren (1983) and Pommerenke (1975), and post-de Branges exposition can be found in Conway (1995).

3.7.1. *Class S*

Bieberbach (1916) proved Theorem 3.3 by giving a sharp upper bound on the second coefficient of a univalent function from class S. Based on this work, he made a conjecture about the other coefficients.

Conjecture 3.35 (Bieberbach). *Let $f(z) = z + \sum_{n=2}^{\infty} a_n z^n$ be a function from S. Then $|a_n| \leq n$. Moreover, $|a_n| = n$ for some n if and only if f is a rotation of the Koebe function.*

The question about the individual coefficients is very subtle, but the growth rate is much easier to estimate. In 1925, Littlewood used integral means of functions from S to prove the following theorem.

Theorem 3.36 (Littlewood). *Let $f(z) = z + \sum_{n=2}^{\infty} a_n z^n$ be a function from the class S. Then $|a_n| \leq en$ for $n \geq 2$.*

This estimate is not sharp, but it gives the optimal growth rate $|a_n| = O(n)$. The proof of this theorem is rather short and elegant, so we provide it here. It is based on a lemma about the growth rate for integral means of functions from the class S.

Lemma 3.37. *Let f be a function from the class S and*

$$M(f,r) = \frac{1}{2\pi}\int_0^{2\pi} |f(re^{i\theta})|\mathrm{d}\theta$$

be its integral mean. Then

$$M(f,r) \leq \frac{r}{1-r}.$$

Proof. Let us consider a function $h(z) = f^{1/2}(z^2) = z + c_2z^2 + \cdots$ which also belongs to the class S. The domain $\Omega_r = h(r\mathbb{D})$ is a bounded domain and its area can be evaluated in terms of the coefficients c_n. Indeed, by the change of variable formula, the area is given by

$$\int_0^r \int_0^{2\pi} s|h'(se^{i\theta})|^2\mathrm{d}\theta\mathrm{d}s = \int_0^r \int_0^{2\pi} sh'\bar{h}'\mathrm{d}\theta\mathrm{d}s.$$

Writing h' and $\bar{h}'$ in terms of the power series, we get

$$\begin{aligned}
&\int_0^r \int_0^{2\pi} s \sum_{n,m=0}^{\infty} c_n\bar{c}_m nms^{n+m-2}e^{i\theta(n-m)}\mathrm{d}\theta\mathrm{d}s \\
&= \int_0^r s2\pi \sum_{n=0}^{\infty} |c_n|^2n^2s^{2n-2}\mathrm{d}s \\
&= \pi\sum_{n=0}^{\infty} |c_n|^2nr^{2n}.
\end{aligned}$$

But by the Growth theorem (Theorem 3.11)

$$|h(re^{i\theta})| = |f(r^2e^{2i\theta})|^{1/2} \leq \left(\frac{r^2}{(1-r^2)^2}\right)^{1/2} = \frac{r}{1-r^2}.$$

This means that the image of $r\mathbb{D}$ is inside the disc of radius $r/(1-r^2)$ and hence

$$\mathrm{Area}(\Omega_r) = \pi\sum_{n=1}^{\infty} |c_n|^2nr^{2n} \leq \pi\frac{r^2}{(1-r^2)^2}.$$

Cancelling out πr, we get

$$\sum_{n=1}^{\infty} n|c_n|^2r^{2n-1} \leq \frac{r}{(1-r^2)^2}.$$

Integrating with respect to r, we arrive at

$$\frac{1}{2}\sum_{n=1}^{\infty}|c_n|^2 r^{2n} = \int_0^r \sum_{n=1}^{\infty} n|c_n|^2 s^{2n-1}\mathrm{d}s$$
$$\leq \int_0^r \frac{s}{(1-s^2)^2}\mathrm{d}s = \frac{1}{2}\frac{r^2}{1-r^2}.$$

The left-hand side can be rewritten as an integral mean of $|h|^2$:

$$\frac{1}{2\pi}\int_0^{2\pi}|h^2(re^{i\theta})|\mathrm{d}\theta = \frac{1}{2\pi}\int_0^{2\pi} h(re^{i\theta})\bar{h}(re^{i\theta})\mathrm{d}\theta = \sum_{n=1}^{\infty}|c_n|^2 r^{2n}.$$

Combining with the previous inequality, we obtain

$$\frac{1}{2\pi}\int_0^{2\pi}|h^2(re^{i\theta})|\mathrm{d}\theta \leq \frac{r^2}{1-r^2},$$

which is equivalent to

$$\frac{1}{2\pi}\int_0^{2\pi}|f(re^{i\theta})|\mathrm{d}\theta \leq \frac{r}{1-r}.$$ □

Proof of Theorem 3.36. By the Cauchy formula,

$$a_n = \frac{1}{2\pi}\int_{|z|=r}\frac{f(z)}{z^{n+1}}\mathrm{d}z, \quad r<1.$$

By the triangle inequality and the previous lemma,

$$|a_n| \leq \frac{1}{r^n}\int_0^{2\pi}|f(re^{i\theta})|\mathrm{d}\theta \leq \frac{1}{r^n}\frac{r}{1-r}.$$

It is easy to check that the right-hand side has the minimal value at $r = 1 - 1/n$. Plugging in this value, we obtain

$$|a_n| \leq \frac{n}{(1-1/n)^{n-1}} = n\left(1+\frac{1}{n-1}\right)^{n-1} < en.$$ □

Later, Bazilevich (1951) improved the upper bound to

$$|a_n| \leq \frac{en}{2} + C$$

for some absolute constant C. These methods were further improved by Baernstein (1974) who proved that $|a_n| \leq en/2$.

An asymptotic version was obtained by Hayman (1955). He proved that

$$\alpha = \lim_{n\to\infty}\frac{|a_n|}{n}$$

exists for all functions $f \in S$. Moreover, $\alpha < 1$ unless f is a rotation of the Koebe function. In particular, this implies that $|a_n| \leq n + 1$ for all sufficiently large n.

The other direction of attack was to prove the Bieberbach conjecture for particular values of n. The case $n = 2$ was proved by Bieberbach himself in 1916. The case $n = 3$ was solved by Loewner in 1923 (Löwner, 1923), who introduced the method of slit domains and Loewner Evolution. Loewner Evolution will be discussed in Chapter 5 and the estimate of the third coefficient will be proved in Section 5.1.3.

The next case $n = 4$ was resolved by Garabedian and Schiffer (1955b). Their proof was very complicated but five years later a much simpler proof was given by Charzynski and Schiffer (1960). A few years later, independently, Pederson (1968/1969) and Ozawa (1969) proved that $|a_6| \leq 6$. The case $n = 5$ turned out to be even more difficult and was proved by Pederson and Schiffer (1972).

Beyond the work on partial cases, there was a lot of work on similar problems. In particular, in 1936, Robertson considered a similar problem for odd functions from S. If $f = \sum a_n z^n$ is from S, then

$$g(z) = f^{1/2}(z^2) = \sum_{k=1}^{\infty} c_k z^{2k-1}$$

is also from S. Robertson (1936) conjectured that

$$\sum_{k=1}^{n} |c_k|^2 \leq n$$

for all $n \geq 2$. It is not so easy to express the coefficients c_n in terms of a_n, but it is easy to verify that

$$a_n = \sum_{k=1}^{n} c_k c_{n+1-k}.$$

By the Cauchy–Schwarz inequality, this implies

$$|a_n| \leq \sum_{k=1}^{n} |c_k|^2.$$

Hence, the Bieberbach conjecture immediately follows from the Robertson conjecture.

In a series of papers in the 1960s, Lebedev and Milin considered the logarithmic coefficients of f, namely, the coefficients of

$$\log \frac{f(z)}{z} = 2\sum_{n=1}^{\infty} \gamma_n z^n.$$

Milin (1977) conjectured that

$$\sum_{k=1}^{n}(n-k+1)\left(k|\gamma_k|^2 - \frac{1}{k}\right) \leq 0$$

and proved that this conjecture implies the Robertson conjecture. The final step was made by de Branges (1985), who used Loewner's techniques together with methods of functional analysis to prove the Milin conjecture and thus both Robertson and Bieberbach conjectures. An interesting explanation of the history of the Bieberbach conjecture, ideas behind de Branges proof, and a historical account about verification of the proof can be found in the paper "The Last 100 Days of the Bieberbach Conjecture" by Fomenko and Kuz'mina (1986) and in the proceedings of the symposium on the occasion of the proof of the Bieberbach conjecture (Drasin *et al.*, 1986). de Branges' proof has been significantly simplified since then. We refer to Conway (1995, Chapter 17) for an exposition of a simplified proof by Fitzgerald and Pommerenke. Eventually, the proof was distilled to just four pages by Weinstein (1991).

3.7.2. *Class* Σ

Surprisingly, a similar question about the coefficients of functions from the class Σ is still open. Moreover, there are no good conjectures about the sharp upper bounds. In fact, even the question about the optimal decay rate of the coefficients is open.

For a univalent function $g(z) = z + \sum b_n z^{-n}$ from class Σ, we know that $|b_n| \leq n^{-1/2}$, but this is not sharp for $n > 1$. For $n = 1$, it is achieved only for $g(z) = z - 1/z$ and its rotation. It is easy to check that

$$g(z) = z - \frac{1}{z} = K^{-1/2}(z^{-2}),$$

where K is the Koebe function. Schiffer (1938) proved that $|b_2| \leq 2/3$. Moreover, the upper bound is sharp and is achieved only by the function

$$K^{-1/3}(z^{-3}) = z - \frac{2}{3}\frac{1}{z^2} + \cdots$$

and its rotations. This led to a natural conjecture that for $|b_n|$, the supremum is equal to $2/(n+1)$ and is achieved by

$$K^{-1/(n+1)}(z^{-(n+1)}) = z - \frac{2}{n+1}\frac{1}{z^n} + \cdots .$$

This very natural conjecture turned out to be wrong. Garabedian and Schiffer (1955a) proved that the sharp upper bound for the third coefficient is

$$|b_3| \le \frac{1}{2} + e^{-6}.$$

The same result for the family of odd functions from Σ was obtained by Bazilevich (1937). These results show that the conjecture $|b_n| \le 2/(n+1)$ fails for $n = 3$. The method of Bazilevich could be used to show that the conjecture fails for all odd n. Moreover, it turned out that the conjecture is wrong even asymptotically. Pommerenke (1967) constructed a function from Σ such that

$$|b_n| \ge n^{0.139-1}.$$

This raises the question: What is the best asymptotic bound on the coefficients? To be more precise, we define γ as the best constant such that

$$|b_n| = O(n^{\gamma+\epsilon-1})$$

for every function $g(z) = z + \sum b_n z^{-n}$ from Σ and every $\epsilon > 0$. A corollary to the Gronwall theorem implies $\gamma \le 1/2$. Clunie and Pommerenke (1967) improved it to $\gamma \le 1/2 - 1/300$. The result of Pommerenke implies that $\gamma \ge 0.139$. Note that the similar constant for the class S is equal to 2, following from Littlewood's estimate $|a_n| \le en$ and the fact that for the Koebe function, $|a_n| = n$.

3.7.3. *Integral means*

Given a close relation between functions from S and Σ, it is puzzling why the coefficient problems for functions from Σ are so different and much harder than their counterparts for functions from S.

A partial answer is that the functions from Σ are more closely related to the *bounded* univalent functions in the unit disc (which we denote by S_b). The boundedness condition changes everything. First, we immediately lose our usual suspect for the extremal function: the Koebe function. Moreover, for a bounded function, the coefficients cannot grow, they must decay. An argument virtually identical to the proof of Gronwall's theorem shows that for a bounded function from S, one has $|a_n| \le n^{1/2}$. This suggests the introduction of an exponent γ_b which is defined as the best constant such that

$$|a_n| = O(n^{\gamma_b+\epsilon-1})$$

for every function $f(z) = z + \sum a_n z^n$ from S_b and every $\epsilon > 0$.

This gives a partial but not quite satisfactory explanation as to why bounded functions are more complicated than the unbounded ones. In the true geometric spirit of this book, the answer comes from the geometric side.

To explain the underlying geometry, we have to consider integral means as in Lemma 3.37, only this time we consider integral means of the derivative. We define

$$I(f,r) = \frac{1}{2\pi}\int_0^{2\pi} |f'(re^{i\theta})|\mathrm{d}\theta.$$

This integral mean has a simple geometric interpretation: up to the factor $2\pi r$, this integral is equal to the length of the image of the circle of radius r under f. Since the circle is the set of points where Green's function in $\mathbb{D}$ is equal to $\log r$ and Green's function is conformally invariant, we have

$$2\pi r I(f,r) = \text{length}\left(\{G_{f(\mathbb{D})}(0,\cdot) = \log r\}\right).$$

This shows that integral means $I(f,r)$ are closely related to the lengths of level lines of G, or, in more physics terms, to the lengths of electrostatic equipotential lines.

It turns out that it is easy to connect the integral means with coefficients via the Cauchy formula. Indeed, we can write $f'(z) = \sum n a_n z^{n-1}$ and by the Cauchy formula for the coefficients of f, we have

$$na_n = \frac{1}{2\pi}\int_{|z|=r} \frac{f'(z)}{z^n}\mathrm{d}z.$$

Changing variables and applying the triangle inequality:

$$|a_n| \le \frac{1}{n}\frac{1}{2\pi}r^{1-n}\int |f'(re^{i\theta})|\mathrm{d}\theta.$$

Plugging $r = 1 - 1/n$ and using $(1-1/n)^{-n} \to e$, we arrive at

$$|a_n| \le \frac{e}{n} I\left(f, 1-\frac{1}{n}\right)(1+o(1)). \tag{3.27}$$

We denote the growth rate of $I(f, 1-1/n)$ by β, that is, β is the best number such that

$$I(f, 1-1/n) = O(n^{\beta+\epsilon})$$

for every f and ϵ. To be completely rigorous, there are two different constants β_S and β_b for classes S and S_b, respectively. The estimate above suggests that $\gamma_b \leq \beta_b$. In fact, it is not too difficult to show that this is indeed true. This computation was essentially done by Littlewood in 1925. Using the same argument but changing $1 - 1/n$ to $1 + 1/n$, one gets an inequality similar to (3.27) for the coefficients b_n of function from Σ, thus showing that $\gamma \leq \beta$, where $\beta = \beta_\Sigma$ is naturally defined as the growth rate of the integral means of g' for $g \in \Sigma$.

Carleson and Jones (1992) proved that in fact $\beta_b = \gamma_b = \beta = \gamma$ and conjectured that $\beta = 1/4$. This is known as the Carleson–Jones conjecture. Their result also explains why the coefficient problems for S_b and Σ are so difficult. In order to have large coefficients, integral means $I(f, r)$ should be large. But I is the length of a curve which approaches the boundary of the image domain. In the unbounded case, these curves could be long because their diameters grow to infinity. This is exactly what happens for the Koebe function. In the cases of S_b and Σ, the boundary of the image is bounded, so for the Green function's equipotential lines to be long, they have to be very 'wiggly', and they should become more and more 'wiggly' as they approach the boundary. This suggests that the boundary should be of fractal nature. Working with conformal maps onto fractal domains is obviously much more difficult than working with maps like the Koebe function. In particular, this explains why it is so difficult to come up with a good candidate for an extremal function: this function should be of 'fractal' nature but such functions are notoriously difficult to write down in an explicit form.

The Carleson–Jones conjecture is still open. The upper bound $\beta \leq 0.46$ was obtained by Hedenmalm and Shimorin (2005). The lower bound $\beta > 0.23$ is due to Beliaev and Smirnov (see Beliaev and Smirnov, 2010; Beliaev, 2008). The cover picture shows a part of the domain which gives the lower bound 0.23.

3.8. Solutions

Solution of 3.1. The function K_α can be written as a composition of three functions that are univalent. The first function $(z+1)/(1-z)$ maps the unit disc onto the right half-plane. After this, we compose it with $z \mapsto z^\alpha$, which maps the right half-plane onto the wedge $\{z : -\pi\alpha/2 < \arg(z) < \pi\alpha/2\}$. Finally, we apply a linear map $z \mapsto (z-1)/2\alpha$. The image of the wedge under this map is also a wedge of the same size and direction, but with the vertex at $-1/2\alpha$. This gives us a description of $K_\alpha(\mathbb{D})$.

Since K_α is a composition of three univalent maps, it is univalent as well. Near the origin, we have

$$\frac{z+1}{1-z} = 1 + 2z + \cdots$$

and

$$\left(\frac{z+1}{1-z}\right)^\alpha = 1 + 2\alpha z + \cdots ,$$

which immediately leads to $K_\alpha(z) = z + \cdots$. □

Solution of 3.2. It is obvious that J_k has the right normalization, so we only have to show that it is univalent. This can be easily shown by considering the equation $z + k/z = w$; for every w, this is a quadratic equation which has two solutions. The product of these solutions is k with $|k| \leq 1$; hence at most one of the roots is from $\mathbb{D}_-$. This proves that J_k is univalent in $\mathbb{D}_-$.

To find the image, it is sufficient to find the image of the boundary. It is easy to compute the image of the unit circle: it is parametrized by $((1+k)\cos(\theta), (1-k)\sin(\theta))$ which is an ellipse with axis along the coordinate axis with semi-axes equal to $(1+k)$ and $(1-k)$. The canonical Joukowsky map corresponds to an extreme case $k = 1$, where the image of the circle is the interval $[-2, 2]$. Thus, the image of $\mathbb{D}_-$ is the exterior of the ellipse (or an interval in the extremal cases $k = 1$ or $k = -1$). □

Solution of 3.3.

(1) Since f_μ is a composition of a Möbius transformation which preserves $\mathbb{D}$ and a linear map, it is well defined and univalent in $\mathbb{D}$. Clearly, $f(\mu(z)) = (f \circ \mu)(0) + (f \circ \mu)'(0)z + \cdots$, which immediately implies that $f_\mu(z) = z + \cdots$.
(2) If $f(z) = z + \sum a_n z^n$, then $\bar{f}(\bar{z}) = z + \sum \bar{a}_n z^n$, which is obviously univalent and has the right normalization.
(3) The function $f(z^n)/z^n$ is analytic and non-vanishing in the unit disc. Hence, we can define a single-valued branch of $(f(z^n)/z^n)^{1/n}$. Let us consider a function

$$g(z) = z\left(\frac{f(z^n)}{z^n}\right)^{1/n}.$$

This function is analytic in $\mathbb{D}$. We claim that it is univalent. Let us assume $g(z_1) = g(z_2)$. Taking power n, we get $f(z_1^n) = f(z_2^n)$ and since f is univalent, $z_1^n = z_2^n$. This is equivalent to $z_2 = z_1 e_n$ where e_n is some root of unity. Plugging this into g, we arrive at

$$z_1 e_n \left(\frac{f(z_1^n)}{z_1^n}\right)^{1/n} = z_1 \left(\frac{f(z_1^n)}{z_1^n}\right)^{1/n},$$

which proves that $e = 1$ and $z_1 = z_2$ or $z_1 = 0$ which also implies $z_1 = z_2$. The fact that g has the right normalization at the origin follows immediately from the definition.

The proofs for the class Σ are essentially the same. Alternatively, it is possible to see that these transforms behave nicely under conjugation by $1/z$. □

Solution of 3.4. Let z_0 be some point in Ω and denote $\operatorname{dist}(z_0, \partial\Omega)$ by r and define

$$g(z) = \frac{f(z_0 + zr)}{rf'(z_0)}.$$

This function is defined in $\mathbb{D}$ and is univalent there. It is easy to check that it is from S. By Corollary 3.5,

$$\begin{aligned}\operatorname{dist}(w_0, \partial\Omega') &= r|f'(z_0)| \operatorname{dist}(0, \partial\Omega'') \\ &\geq r|f'(z_0)| \operatorname{dist}(0, \partial g(\mathbb{D})) \geq \frac{r|f'(z_0)|}{4},\end{aligned}$$

where $w_0 = f(z_0)$ and $\Omega'' = (\Omega' - w_0)/r|f'(z_0)|$. This proves the second inequality in (3.3). Applying the same argument to $f^{-1} : \Omega' \to \Omega$, we obtain the first inequality. □

Solution of 3.5. Function f maps the unit disc onto a bounded domain Ω. Let z_n be a sequence in $\mathbb{D}$ with $|z_n| \to 1$. By the Koebe distortion theorem (Theorem 3.7),

$$(1 - |z_n|)|f'(z_n)| \leq 4 \operatorname{dist}(f(z_n), \partial\Omega).$$

By the boundary correspondence (see Proposition 2.18), $f(z_n)$ must tend to the boundary of Ω; since Ω is bounded, the distance from $f(z_n)$ to the boundary must tend to 0.

To show that the boundedness is necessary, we consider the usual suspect: the Koebe function $K(z) = z/(1-z)^2$. Direct computation shows that $K'(z) \approx 2/(1-z)^3$ as $z \to 1$. □

Solution of 3.6. Let

$$\phi(z)=\frac{f(\frac{z+w_0}{1+\bar{w}_0 z})-f(w_0)}{(1-|w_0|^2)f'(w_0)}$$

be the Koebe transform of f with respect to z_0 as in (3.1). From the formula for ϕ and the fact that the ranges of f and $f((z+z_0)/(1+\bar{z}_0 z))$ are the same, we obtain

$$\operatorname{dist}(f(z_0),\partial f(\mathbb{D}))=(1-|z_0|^2)|f'(z_0)|\operatorname{dist}(\phi(0),\partial\phi(\mathbb{D})).$$

By Exercise 3.3, ϕ is in S, so we can apply Corollary 3.5:

$$\frac{1}{4}\le\operatorname{dist}(0,\partial\phi(\mathbb{D}))\le 1.$$

Together with the previous formula, this completes the proof. □

Solution of 3.7. Let us define $h(w)$ to be the Koebe transform of f with respect to z (see formula (3.5)). It is clear that

$$h(-z)=\frac{-f(z)}{(1-|z|^2)f'(z)}.$$

Since $h\in S$, we can apply the Growth theorem to h to obtain

$$\frac{r}{(1+r)^2}\le\frac{|f(z)|}{(1-r^2)|f'(z)|}\le\frac{r}{(1-r)^2},$$

which is equivalent to (3.10). □

Solution of 3.8. The solution is based on the duality between functions from S and Σ'. Let us denote $w=1/z$ and consider a function $h(w)=1/f(w)$ which belongs to the class S. By the Growth theorem,

$$\frac{|w|}{(1+|w|)^2}\le|h(w)|\le\frac{|w|}{(1-|w|)^2}.$$

Expressing h and w in terms of f and z and reversing the inequalities, we obtain the desired estimate on $|f|$. □

Solution of 3.9. Let K be compact. Since the conformal radius of Ω_n tends to infinity, by Corollary 3.6, the inradius of Ω_n also tends to infinity and $K \subset \Omega_n$ for sufficiently large n. This proves that g_n is defined on K for large n. Let us consider a function $|f_n'(0)|^{-1}f_n(z)$ which belongs to the class S. Applying the Growth theorem, we have

$$\frac{|z|}{(1+|z|)^2} \le |f_n'(0)|^{-1}|f_n(z)| \le \frac{|z|}{(1-|z|)^2}.$$

Denoting $w = f_n(z)$, these inequalities can be rearranged into

$$(1-|g_n(w)|)^2|w| \le |f_n'(0)||g_n(w)| \le (1+|g_n(w)|)^2|w|.$$

Since $|g_n| \le 1$, the second inequality implies that $|g_n(w)| \to 0$ uniformly on compacts. This, together with the double inequality above, implies that

$$\frac{|f_n'(0)||g_n(w)|}{|w|} \to 1$$

uniformly on compacts. Hence, the family of maps $|f_n'(0)|g_n(w)/w$ is normal. By normality, from every subsequence, we can choose a converging sub-subsequence. The limit has modulus 1 everywhere and equal to 1 at the origin; hence it is 1 everywhere. Since the limit is the same for all subsequences, the sequence $|f_n'(0)|g_n(w)/w \to 1$ uniformly on compacts.

The second part follows immediately by applying the result of the first part to the function $\tilde{g}_n(z) = 1/g(1/z)$. □

Solution of 3.10. Let Ω be the kernel of (Ω_n) as defined by Definition 3.13 and $\tilde{\Omega}$ be the kernel defined by Definition 3.14. From now on, we assume that both kernels are non-trivial. For every point $w \in \tilde{\Omega}$, $\tilde{\Omega}$ is a connected open domain containing w_0 and w; hence $w \in \Omega$. Equivalently, $\tilde{\Omega} \subset \Omega$.

Let us assume that $\tilde{\Omega} \neq \Omega$, that is, there is $w \in \Omega \setminus \tilde{\Omega}$. Let U be an open connected domain containing w and w_0 and contained in almost all Ω_n. Then the domain $\tilde{\Omega} \cup U$ is a domain containing w_0 such that every closed subset belongs to almost all Ω_n. Moreover, by our assumption, this domain is strictly larger than $\tilde{\Omega}$, which is impossible by the definition of $\tilde{\Omega}$. This completes the proof that $\Omega = \tilde{\Omega}$. □

Solution of 3.11. Let us first assume that w_0 is on the boundary of Ω and that the kernel is non-trivial. In this case, there is another point in the kernel, and hence there is an open domain containing w_0 and contained in almost all domains Ω_n. Since the domains are decreasing, this open domain is a subdomain of all Ω_n and hence w_0 is an interior point of Ω. This contradicts our assumption and proves that the kernel must be trivial.

Now, we assume that w_0 is an interior point of Ω and denote the component of interior of Ω containing w_0 by D. First, it is easy to see that D

is inside the kernel. Indeed, for every point of D there is an open domain (D itself) which contains this point and w_0 and contained in all Ω_n. The opposite inclusion is also simple: let w be a point in the kernel, then there is an open domain containing both w_0 and w and contained in all Ω_n, and, hence in Ω. An open connected subdomain of Ω must be inside one of the components of the interior. Since it contains w_0, it must be inside D. In particular, $w \in D$. □

Solution of 3.12. It is clear that the kernel for every $t_n \to \pi$ is the complement of the unit disc, which proves the convergence of domains. By (3.28),

$$g_t(z) = \frac{(z-1)^2 + 2z\cos^2(t/2) + (z+1)\sqrt{(z+1)^2 - 4z\cos^2(t/2)}}{2z}.$$

When z is in any closed set away from $\mathbb{D}$, as $t \to \pi$, this function converges uniformly to

$$\frac{(z-1)^2 + (z+1)\sqrt{(z+1)^2}}{2z} = z + \frac{1}{z}. \qquad \square$$

Solution of 3.13. By Exercise 2.6, $\mathcal{P}$ is normal, that is, from each sequence, one can choose a convergent subsequence. Since the uniform limit of analytic functions is analytic and conditions (3.14) are preserved by limits, the limit is also an element of $\mathcal{P}$. This proves that $\mathcal{P}$ is compact. □

Solution of 3.14. It is easy to find the uniformizing map g_K. Since the Joukowsky map $z + 1/z$ maps the complement of the unit disc onto the plane with the slit $[-2, 2]$, the map $z/R + R/z$ sends Ω onto $\mathbb{H}$. By one more rescaling, we obtain $g_K = z + R^2/z$ and so $\text{hcap}(K) = R^2$. □

Solution of 3.15. In Exercise 2.8, we found the uniformizing map. The standard solution gives $g(z) = (z^2 + t^2)^{1/2}$, where we use the branch of the square root which has a positive imaginary part. This function sends infinity to infinity, so we only have to find its expansion at infinity. Simple computation shows that

$$g(z) = z(1 + t^2/z^2)^{1/2} = z + \frac{t^2}{2z} + \cdots .$$

This proves that $g = g_K$ and that $\text{hcap}(K) = t^2/2$. □

Solution of 3.16. Both parts are proved by considering the corresponding uniformizing maps. In the first part, it is easy to see that $g_K(z - x) + x$

maps the complement of $K + x$ onto $\mathbb{H}$. Expanding at infinity, we get

$$\begin{aligned} g_K(z - x) + x &= z - x + x + \frac{\text{hcap}(K)}{z - x} + \cdots \\ &= z + \frac{\text{hcap}(K)}{z} + \cdots, \end{aligned}$$

which proves that $\text{hcap}(K) = \text{hcap}(K + x)$.

To prove the second part, we observe that $g_K(g_{K_1}(z))$ maps the complement of K_2 onto $\mathbb{H}$. Combining series expansions for g_K and g_{K_1} we obtain

$$g_K(g_{K_1}(z)) = z + \frac{\text{hcap}(K) + \text{hcap}(K_1)}{z} + \cdots.$$

By uniqueness of the Riemann map with thermodynamic normalization, we see that $g_K(g_{K_1}) = g_{K_2}$ and $K = g_{K_1}(K_2 \setminus K_1)$. □

Solution of 3.17. There are several ways to show that f is a conformal map from $\mathbb{H}$ onto $\mathbb{H}$ with a slit. One approach is to show directly that f is injective and then directly compute the image of the real line.

We will follow a different approach. The derivative of f is

$$f'(z) = (z + \alpha)^{-\alpha}(z - (1 - 2\alpha))(z - (1 - \alpha))^{\alpha - 1}.$$

This has the form of derivative of the Christoffel–Schwarz map (see Theorem 2.17), and hence f maps the upper half-plane onto a polygonal domain. Points $-\alpha$, $1 - 2\alpha$, and $1 - \alpha$ are mapped to the corners of this polygonal domain with angles $(1 - \alpha)\pi$, 2π, and $\alpha\pi$. Since both $-\alpha$ and $1 - \alpha$ are mapped to the origin, the image of $\mathbb{H}$ is the upper half-plane with a straight slit from 0 to $f(1 - 2\alpha) = (1 - \alpha)^{1-\alpha}\alpha^{\alpha}e^{i\alpha\pi}$.

To compute the capacity, we expand f at infinity to find

$$f(z) = z - \frac{\alpha(1 - \alpha)}{2z} + \cdots,$$

which means that $\text{hcap}(K) = \alpha(1 - \alpha)/2$. Rescaling K to be the unit interval and using the scaling property of capacity, we obtain the formula for $\text{hcap}(I_\alpha)$. □

Solution of 3.18. First, we apply $(z - 1)/(z + 1)$ to send the unit circle to the imaginary line. Under this map, the tip $e^{i\theta}$ is mapped to iy_θ, where $y_\theta = \tan(\theta/2)$ and the image of the domain is the upper half-plane with the vertical slit from 0 to iy_θ. We already know that this domain is mapped out using $(z^2 + y_\theta^2)^{1/2}$. Under this map, points ± 1 are mapped to $\pm x_\theta$, where $x_\theta = 1/\cos(\theta/2)$. Finally, we apply one more Möbius transformation which

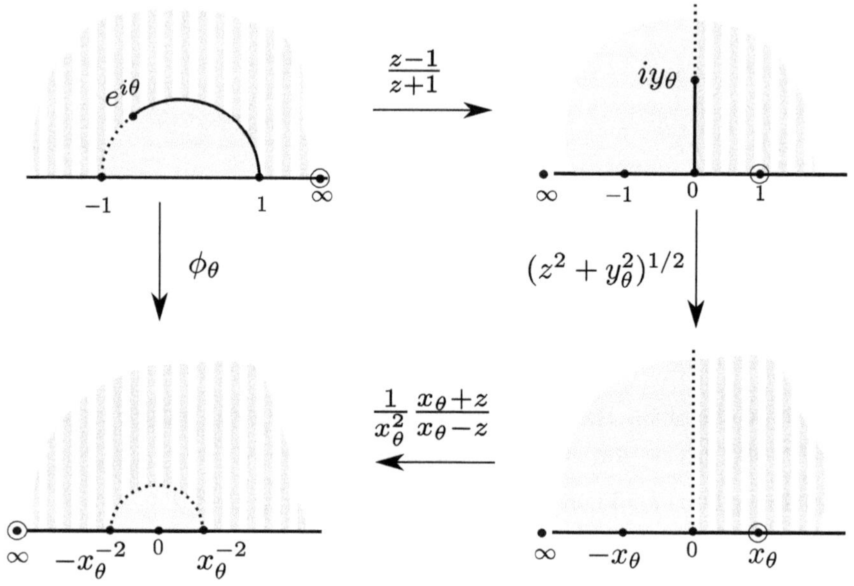

Fig. 3.5. Mapping out a circular slit. The dashed line is the part of the circle which is not mapped out and all of its images. To trace how the functions act, we use different backgrounds for the inside of the unit disc and its images and for the outside. The images of infinity are marked by a double circle.

sends x_θ, the image of infinity, back to infinity. We define ϕ_θ to be the composition of these three maps (see Figure 3.5). Expanding all functions at the corresponding points up to terms of order -2, we can get the first few terms of the power series of ϕ_θ at infinity:

$$\phi_\theta(z) = z + 2\sin^2(\theta/2) + \frac{1-\cos^4(\theta/2)}{z} + O\left(\frac{1}{z^2}\right).$$

The normalized map is

$$g_\theta(z) = \phi_\theta(z) - 2\sin^2(\theta/2)$$

and

$$\operatorname{hcap}(K_\theta) = 1 - \cos^4(\theta/2).$$

The function ϕ_θ can be written in a more explicit form as

$$\frac{(z-1)^2 + 2z\cos^2(\theta/2) + (z+1)\sqrt{(z+1)^2 - 4z\cos^2(\theta/2)}}{2z}. \tag{3.28}$$

As before, we use the branch of the square root which has positive *imaginary* part. □

Solution of 3.19. Let us consider the inverse functions f_K and apply formula (3.21), so that

$$f_K(x) - x = \frac{1}{\pi} \int_I \frac{\operatorname{Im} f_K(t)}{t - x} \mathrm{d}t,$$

where I is the set of points with $\operatorname{Im} f_K > 0$, that is, the image of ∂K under g_K. For points x that are to the right of I, the integral is obviously negative since $\operatorname{Im} f_K > 0$ and $t - x < 0$. This proves that $f_K(x) - x < 0$ or, equivalently, $x - g_K(x) < 0$. For points that are to the left of I, the argument is exactly the same up to the change of sign. □

Chapter 4

Extremal Length and Other Conformal Invariants

In this chapter, we discuss various quantities that do not change under conformal transformations or change in a very simple, predictable, and controllable way.

We have seen examples of conformal invariants. The first class of examples stemmed from the fact that harmonic functions are conformally invariant. This implied that solutions to the Dirichlet boundary problem are also conformally invariant. In particular, this gives the conformal invariance of Green's function and harmonic measures. These are important examples of *functions* that are conformally invariant. There are also some *quantities* that are conformally invariant. We have already encountered two examples of this type: two rectangles or two annuli are conformally equivalent if and only if they have the same 'shape'. More generally, given a conformal rectangle or a doubly connected domain, they can be mapped onto a proper rectangle or an annulus, respectively. The 'shape' of these uniformizing domains is a conformal invariant. This approach works with virtually any definition of the shape. In the case of rectangles, the canonical notion is the ratio of side lengths and in the case of annuli it is the conformal modulus. In this chapter, we develop a general and versatile approach to conformal invariants. In particular, we see that the 'shape' of a rectangle and an annulus are particular instances of the same notion.

4.1. Extremal Length

Extremal length is a conformal invariant which has a simple geometric definition; this makes it a very powerful tool. Here, we discuss the main results and applications of extremal length. More information can be found in Ahlfors (2010), Beurling (1989), Goluzin (1969), Garnett and Marshall (2008) and Ohtsuka (1970).

The introduction of extremal length is frequently attributed to Ahlfors; in fact, in its modern form, it was introduced by Beurling in the early 1940s and later developed by Beurling and Ahlfors. Some of the underlying ideas can be traced back to the work of Grötzsch. As Ahlfors and Beurling stated (see Beurling (1989, p. 213)), "Historically our method has its roots in previous individual work by both of us. It is also, in special cases, closely related to the 'method of strips' of Grötzsch. We believe, however, that it differs fundamentally from the method of Grötzsch by virtue of its generality, flexibility, and extreme simplicity." These are exactly the reasons why this type of conformal invariant is particularly useful.

4.1.1. *Definitions and basic properties*

Let Ω be a domain in $\mathbb{C}$. In this section, we are interested in various collections of curves γ in Ω. Abusing notation, by *curve* we mean a finite (or countable) union of rectifiable arcs in Ω. A *metric* in Ω is a non-negative Borel measurable function ρ such that the *area* of Ω, which is defined as

$$A(\Omega, \rho) = \int_{\Omega} \rho^2(z)\mathrm{d}m(z),$$

satisfies $0 < A(\Omega, \rho) < \infty$. Such metrics are called *admissible*.

Given a metric ρ, we can define the *length* of any rectifiable curve γ as

$$L(\gamma, \rho) = \int_{\gamma} \rho(s)|\mathrm{d}z| = \int_{\gamma} \rho(s)\mathrm{d}s,$$

where $\mathrm{d}s$ is the usual arc length. For a family of curves Γ, we define the minimal length by

$$L(\Gamma, \rho) = \inf_{\gamma \in \Gamma} L(\gamma, \rho).$$

Definition 4.1. Let Γ be a family of rectifiable curves in a domain Ω. The *extremal length* of Γ is defined as

$$\lambda(\Gamma) = \lambda_{\Omega}(\Gamma) = \sup_{\rho} \frac{L^2(\Gamma, \rho)}{A(\Omega, \rho)},$$

where the supremum is over all admissible metrics. The *extremal metric* is a metric for which the supremum is achieved.

The expression in the definition of the extremal length is obviously homogeneous with respect to ρ. This means that we can normalize ρ by fixing $L(\Gamma, \rho)$ or $A(\Omega, \rho)$ or any linear relation between them. Indeed, by rescaling ρ, one can see that

$$\lambda_\Omega(\Gamma) = \sup_\rho L^2(\Gamma, \rho),$$

where the supremum is over all metrics with $A(\Omega, \rho) = 1$. Alternatively,

$$\frac{1}{\lambda_\Omega(\Gamma)} = \inf_\rho A(\Omega, \rho),$$

where the infimum is over all metrics with $L(\Gamma, \rho) = 1$. The quantity $m_\Omega(\Gamma) = \lambda_\Omega(\Gamma)^{-1}$ is called the *(conformal) modulus* of Γ. Finally,

$$\lambda_\Omega(\Gamma) = \sup_\rho L(\Gamma, \rho) = \sup_\rho A(\Omega, \rho),$$

where the supremum is over metrics with $L(\Gamma, \rho) = A(\Omega, \rho)$.

The main property of extremal length is its conformal invariance.

Theorem 4.2. *Let $f : \Omega \to \Omega'$ be a conformal map, and let Γ and Γ' be two families of curves in Ω and Ω' such that $\Gamma' = f(\Gamma)$. Then $\lambda_\Omega(\Gamma) = \lambda_{\Omega'}(\Gamma')$.*

Proof. Let ρ' be a metric in Ω'. Then $\rho(z) = |f'(z)|\rho'(f(z))$ is a metric in Ω and by the change of variable formula $A(\Omega, \rho) = A(\Omega', \rho')$. By the same argument, if $\gamma' = f(\gamma)$, then $L(\gamma, \rho) = L(\gamma', \rho')$. This proves that for every metric ρ', there is a metric ρ such that

$$\frac{L^2(\Gamma, \rho)}{A(\Omega, \rho)} = \frac{L^2(\Gamma', \rho')}{A(\Omega', \rho')}.$$

This implies that $\lambda_\Omega(\Gamma) \geq \lambda_{\Omega'}(\Gamma')$. Applying the same argument to f^{-1}, we complete the proof of the theorem. □

It is also important to note that the definition of extremal length depends on Γ but not on Ω, namely, if we have two domains $\Omega \subset \Omega'$ and Γ is a family of curves in Ω, then $\lambda_\Omega(\Gamma) = \lambda_{\Omega'}(\Gamma)$. The proof of this independence is quite simple. Let ρ be some metric in Ω; we can extend it to ρ' in Ω' by setting $\rho' = 0$ outside of Ω. Obviously, areas and lengths for these two measures are the same and we have $\lambda_\Omega(\Gamma) \leq \lambda_{\Omega'}(\Gamma)$. For any ρ' in Ω', we define ρ to be its restriction to Ω. Clearly, $L(\Gamma, \rho) = L(\Gamma, \rho')$ and

$A(\Omega,\rho) \le A(\Omega',\rho')$, and this implies the opposite inequality:

$$\lambda_{\Omega'}(\Gamma) = \sup_{\rho'} \frac{L^2(\Gamma,\rho')}{A(\Omega',\rho')} \le \sup_{\rho'} \frac{L^2(\Gamma,\rho)}{A(\Omega,\rho)} \le \lambda_\Omega(\Gamma).$$

4.1.2. *Extremal metric*

In general, we do not know which families Γ admit an extremal metric, but it is not difficult to show that if it does exist, then it is essentially unique.

Theorem 4.3. *Let Γ be a family of curves in Ω and let ρ_1 and ρ_2 be two extremal metrics normalized by $A(\Omega,\rho_i) = 1$. Then $\rho_1 = \rho_2$ almost everywhere.*

Proof. For these two metrics, we have that $\lambda(\Gamma) = L^2(\Gamma,\rho_i)$. Let us consider a metric $\rho = (\rho_1+\rho_2)/2$. Then

$$L(\Gamma,\rho) = \inf_\gamma \int_\gamma \frac{\rho_1(z)+\rho_2(z)}{2}|dz| \ge \frac{L(\Gamma,\rho_1)+L(\Gamma,\rho_2)}{2} = \lambda^{1/2}(\Gamma).$$

By the Cauchy–Schwarz inequality,

$$\begin{aligned} A(\Omega,\rho) &= \int_\Omega \frac{(\rho_1+\rho_2)^2}{4} = \frac{A(\Omega,\rho_1)}{4} + \frac{A(\Omega,\rho_2)}{4} + \int_\Omega \frac{\rho_1\rho_2}{2} \\ &\le \frac{1}{2} + \frac{1}{2}\left(\int_\Omega \rho_1^2\right)^{1/2}\left(\int_\Omega \rho_2^2\right)^{1/2} = 1. \end{aligned} \tag{4.1}$$

Together with the previous formula, this implies that

$$\frac{L^2(\Gamma,\rho)}{A(\Omega,\rho)} \ge \lambda(\Gamma).$$

By the definition of extremal length, this must be an equality, ρ must be also an extremal metric and we must have an equality in (4.1). We know that equality occurs in the Cauchy–Schwarz inequality if and only if ρ_1 and ρ_2 are proportional to each other almost everywhere. The normalization $A(\Omega,\rho_1) = A(\Omega,\rho_2)$ implies that they must be equal almost everywhere. □

As we will see in Section 4.1.3, the computation of the extremal length quite often involves making a good guess for the extremal metric. This can be done in surprisingly many cases, but not always. Given a metric, one can quite often show that it is indeed an extremal metric by using a condition that was proposed by Beurling.

Theorem 4.4. *A metric ρ_0 is extremal for a curve family Γ in Ω if there is a subfamily Γ_0 such that*

$$\int_\gamma \rho_0(s)\mathrm{d}s = L(\Gamma, \rho_0), \qquad \text{for all } \gamma \in \Gamma_0$$

and for all real-valued measurable h in Ω, we have that $\int_\Omega h\rho_0 \geq 0$ if $\int_\gamma h\mathrm{d}s \geq 0$ for all $\gamma \in \Gamma_0$.

One can consider Γ_0 as a collection of the shortest curves in Γ that covers the entire support of ρ_0. This theorem is more useful than one might think. It is true that in most cases there is no way to guess the extremal metric; moreover, it might not exist or it might be that it cannot be written in an explicit form. But we might be able to exploit the conformal invariance. In many cases, it is possible to find a conformally equivalent configuration where the extremal metric is particularly simple. This method surprisingly works often. We use it several times throughout this book.

Proof of Theorem 4.4. Let ρ be some other metric normalized by $L(\Gamma, \rho) = L(\Gamma, \rho_0)$. Since all curves from Γ_0 have minimal length with respect to ρ_0, we have that $L(\gamma_0, \rho) \geq L(\gamma_0, \rho_0)$ for any $\gamma_0 \in \Gamma_0$. This implies that for $h = \rho - \rho_0$

$$\int_{\gamma_0} h(s)\mathrm{d}s \geq 0, \qquad \text{for all } \gamma_0 \in \Gamma_0.$$

By assumptions, this implies that

$$\int_\Omega (\rho(z) - \rho_0(z))\rho_0(z)\mathrm{d}x\mathrm{d}y = \int_\Omega h(z)\rho_0(z)\mathrm{d}x\mathrm{d}y \geq 0.$$

This inequality, together with the Cauchy–Schwarz inequality, gives

$$\int_\Omega \rho_0^2 \leq \int_\Omega \rho\rho_0 \leq \left(\int_\Omega \rho^2\right)^{1/2} \left(\int_\Omega \rho_0^2\right)^{1/2}$$

and

$$A(\Omega, \rho_0) = \int_\Omega \rho_0^2 \leq \int_\Omega \rho^2 = A(\Omega, \rho).$$

The last inequality, together with the normalization of ρ, proves that ρ_0 is extremal. $\square$

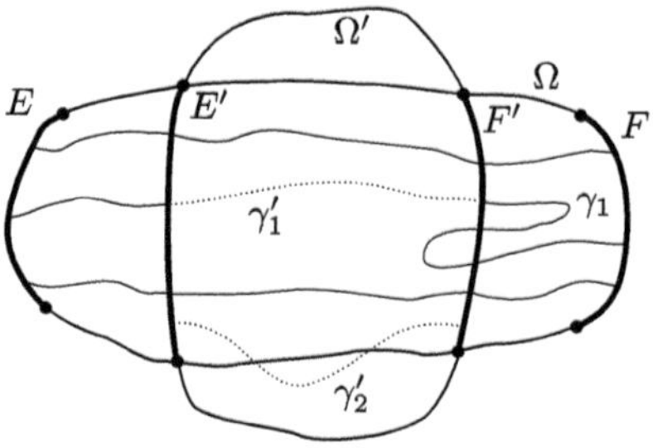

Fig. 4.1. Γ and Γ' are the families of curves connecting E to F and E' to F' within Ω and Ω', respectively. Each curve from Γ contains a dotted piece which belongs to Γ'. The curve $\gamma_2' \in \Gamma'$ is not a part of any curve from Γ.

4.1.3. *Composition rules*

Although it might be hard to compute extremal length in many instances, there are some simple 'rules' that allow us to compare them, which is sufficient for many purposes.

Proposition 4.5 (The comparison rule). *Extremal length is monotone, namely, let Γ and Γ' be two families of curves such that each curve $\gamma \in \Gamma$ contains a curve $\gamma' \in \Gamma'$; then $\lambda(\Gamma) \geq \lambda(\Gamma')$. In other words, a smaller family of longer curves has greater extremal length (see Figure 4.1).*

The proof of this statement is really trivial: just by the definition, $L(\Gamma, \rho) \geq L(\Gamma', \rho)$ and the admissible metrics are the same.

Proposition 4.6 (The serial rule). *Let Ω_1 and Ω_2 be two disjoint domains and Γ_i be two families of curves in these domains. Let Ω be a third domain such that $\Omega_i \subset \Omega$ and Γ be a family of curves in Ω such that each $\gamma \in \Gamma$ contains a curve from each Γ_i (see Figure 4.2 for a typical example). Then $\lambda(\Gamma) \geq \lambda(\Gamma_1) + \lambda(\Gamma_2)$.*

Proof. If any of $\lambda(\Gamma_i)$ is trivial, that is, equal to 0 or ∞, then the statement follows immediately from the comparison rule. From now on, assume that both lengths are non-trivial. Let ρ_i be two metrics normalized by $A(\Omega_i, \rho_i) = L(\Gamma_i, \rho_i)$ and define ρ to be ρ_i in Ω_i and 0 everywhere else. For this metric in Ω we have

$$L(\Gamma, \rho) \geq L(\Gamma_1, \rho_1) + L(\Gamma_2, \rho_2)$$

and

$$A(\Omega, \rho) = A(\Omega_1, \rho_1) + A(\Omega_2, \rho_2) = L(\Gamma_1, \rho_1) + L(\Gamma_2, \rho_2).$$

Combining these two equations, we have $\lambda(\Gamma) \geq \lambda(\Gamma_1) + \lambda(\Gamma_2)$. □

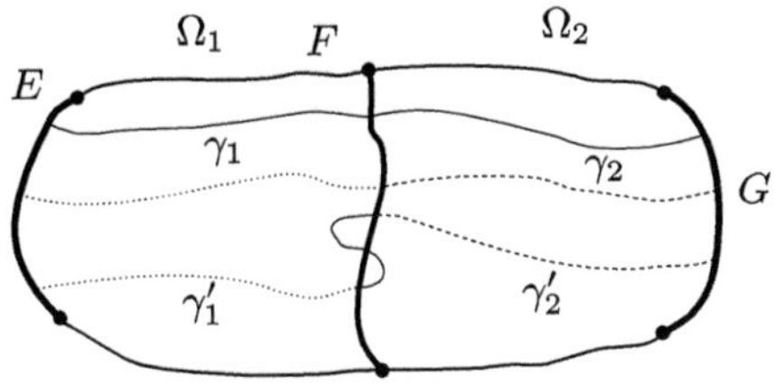

Fig. 4.2. Γ_1 is the family of curves connecting E and F in Ω_1; Γ_2 connects F and G in Ω_2, and Γ connects E and G in $\Omega = \Omega_1 \cup \Omega_2$. Each curve from Γ contains a dotted piece from Γ_1 and dashed piece from Γ_2.

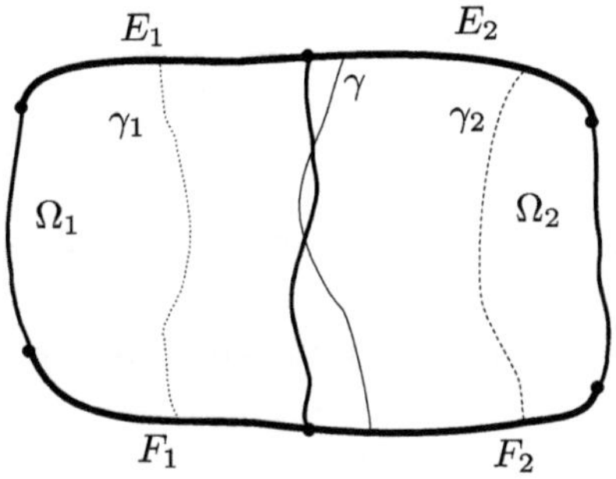

Fig. 4.3. For $i = 1, 2$, Γ_i is the family of curves connecting E_i and F_i inside Ω_i, Γ is the family of curves connecting $E = E_1 \cup E_2$ and $F = F_1 \cup F_2$ inside Ω which is the interior of the closure of $\Omega_1 \cup \Omega_2$. Each curve from Γ_i contains a curve from Γ. In fact, they belong to Γ, but there are curves like $\gamma \in \Gamma$ that are not related to any curve from Γ_i.

Proposition 4.7 (The parallel rule). *Let Ω_1 and Ω_2 be two disjoint domains and Γ_i for $i = 1, 2$ be families of curves in these domains. Let Γ be a third family of curves such that every curve $\gamma_i \in \Gamma_i$ contains a curve $\gamma \in \Gamma$ (see Figure 4.3 for a typical example). Then*

$$\frac{1}{\lambda(\Gamma)} \geq \frac{1}{\lambda(\Gamma_1)} + \frac{1}{\lambda(\Gamma_2)}.$$

Equivalently,

$$m(\Gamma) \geq m(\Gamma_1) + m(\Gamma_2),$$

where m is the conformal modulus.

Proof. Let Ω be a domain containing Γ. Consider a metric ρ in Ω normalized by $L(\Gamma, \rho) = 1$. Our assumptions immediately imply that

$L(\Gamma_i, \rho) \geq L(\Gamma, \rho) = 1$ and

$$A(\Omega, \rho) \geq A(\Omega_1, \rho) + A(\Omega_2, \rho) \geq \frac{1}{\lambda(\Gamma_1)} + \frac{1}{\lambda(\Gamma_2)},$$

where the last inequality follows from $1/A(\Omega_i, \rho) \leq L^2(\Gamma_i, \rho)/A(\Omega_i, \rho) \leq \lambda(\Gamma)$. On the other hand, $\inf A(\Omega, \rho) = 1/\lambda(\Gamma)$ where the infimum is over all metrics normalized by $L(\Gamma, \rho) = 1$. □

Proposition 4.8 (The symmetry rule). *Let Ω be a domain symmetric with respect to the real line and Γ a symmetric family of curves. This means that for every curve $\gamma \in \Gamma$, its symmetric image $\bar{\gamma}$ is also from Γ. Then*

$$\lambda(\Gamma) = \sup_{\rho} \frac{L^2(\Gamma, \rho)}{A(\Omega, \rho)},$$

where the supremum is over all symmetric metrics ρ such that $\rho(z) = \rho(\bar{z})$.

Proof. The proof is almost trivial. Let ρ_1 be an admissible metric and let $\rho_2(z) = \rho_1(\bar{z})$ be its symmetric image. Obviously, $L(\Gamma, \rho_1) = L(\Gamma, \rho_2)$ and $A(\Omega, \rho_1) = A(\Omega, \rho_2)$. By the same argument as in the proof of Theorem 4.3, we have

$$\frac{L^2(\Gamma, \rho)}{A(\Omega, \rho)} \geq \frac{L^2(\Gamma, \rho_1)}{A(\Omega, \rho_1)} = \frac{L^2(\Gamma, \rho_2)}{A(\Omega, \rho_2)},$$

where $\rho = (\rho_1 + \rho_2)/2$. This proves that the supremum over symmetric metrics is equal to the supremum over all admissible metrics. □

Exercise 4.1. State and prove a version of the symmetry rule for symmetry with respect to the unit circle.

4.2. Applications

4.2.1. *Examples and further properties*

There are several configurations of domains with some marked points that are particularly simple. They are either all conformally invariant or defined by a single conformally invariant parameter: a simply connected domain with four marked points on the boundary (conformal rectangle), a simply connected domain with a marked point inside and two marked points on the boundary, a simply connected domain with two marked interior points, or a doubly connected domain. In each of these cases, we already know a conformal invariant that defines the conformal type of the configuration.

In the first case, this is the modulus of a rectangle; in the second case, it is the harmonic measure of an arc between two boundary points evaluated at the interior point; in the third case, it is Green's function; and, in the last case, it is the conformal modulus of the domain. It turns out that all of these invariants can be expressed in terms of extremal lengths.

One of the most important examples of extremal length is the *extremal distance*. Let E and F be two subsets of the closure of Ω. Then the extremal distance between them inside Ω is

$$d_\Omega(E, F) = \lambda(\Gamma),$$

where Γ is the family of all rectifiable curves in Ω that connect E and F. The *conjugate extremal distance* is

$$d^*_\Omega(E, F) = \lambda(\Gamma^*),$$

where Γ^* is the family of all (not necessary connected) curves separating E and F inside Ω. Proposition 4.5 immediately implies that $d_\Omega(E, F)$ decreases when any of Ω, E, or F increases. Figures 4.2 and 4.3 give examples of how Propositions 4.6 and 4.7 could be applied to extremal distances.

Conformal rectangle. Let Ω be a simply connected domain with four marked (accessible) points on the boundary. These points divide the boundary into four connected pieces (again defined in terms of accessible points or prime ends). Let us choose two of them that do not share a common marked point and call them E and F. We know that there is a map from Ω onto a rectangle $[0, a] \times [0, b]$ with some a and b such that the four marked points are mapped to the vertices. Let us assume that the images of E and F are the vertical edges. We claim that the extremal distance $d_\Omega(E, F) = a/b$, which is the conformal invariant that we have seen before.

By conformal invariance of the extremal length, $d_\Omega(E, F)$ is the same as the extremal distance between vertical sides of the rectangle $R = \{(x, y) : 0 < x < a,\ 0 < y < b\}$. For $\rho = 1$, we have that $A(R, \rho) = ab$ and $L(\Gamma, \rho) = a$, where Γ is the family of all curves connecting two vertical sides. This immediately gives us that $\lambda_R(\Gamma) \geq a^2/ab = a/b$.

We claim that this metric is extremal and $\lambda_R(\Gamma) = a/b$. Let Γ_0 be the family of all horizontal lines connecting two vertical sides. Clearly, these curves have the same length, which is equal to $L(\Gamma, \rho)$. If, for some function

h, we have that

$$\int h(x,y)\mathrm{d}x \geq 0, \quad \text{for all } y,$$

then integrating with respect to y, we get

$$\int_R h\mathrm{d}x\mathrm{d}y \geq 0.$$

By Theorem 4.4, this implies that $\rho = 1$ is indeed an extremal metric.

By the same argument, we see that the extremal distance between the other two parts of the boundary is given by b/a and is equal to $d^*_\Omega(E,F)$. In particular, we see that

$$d^*_\Omega(E,F)d_\Omega(E,F) = 1.$$

Exercise 4.2. Use the extremal length to show that the conformal rectangle which is the upper half-plane with marked points -1, 0, 1, and ∞ is conformally equivalent to a square.

Exercise 4.3. Let Ω_1 be a domain in the upper half-plane and let E_1 and F_1 be two disjoint sets on $\partial\Omega$. Let Ω_2, E_2, and F_2 be their symmetric images with respect to $\mathbb{R}$. We define $\Omega = \Omega_1 \cup \Omega_2$ (to be completely rigorous, we also have to add the real part of the boundary), $E = E_1 \cup E_2$, and $F = F_1 \cup F_2$. Show that

$$2d_\Omega(E,F) = d_{\Omega_1}(E_1,F_1) = d_{\Omega_2}(E_2,F_2).$$

Exercise 4.4. Let $0 < \varepsilon < 1$ and let $\Omega = [-1,1]\times[-1,1]\setminus(-\varepsilon,\varepsilon)\times(-\varepsilon,\varepsilon)$ be the 2×2 square without central $2\varepsilon \times 2\varepsilon$ square. Let Γ be the family of curves connecting the top and bottom (that is, $[-1,1]\times\{1\}$ and $[-1,1]\times\{-1\}$) inside Ω. Show that

$$\frac{1}{1-\varepsilon^2} \leq \lambda(\Gamma) \leq \frac{1}{1-\varepsilon}.$$

Conformal annulus. A doubly connected domain is characterized by a single conformal invariant: its modulus. It turns out that this is equal to the extremal distance between the boundary components. This gives a more geometrical interpretation of the conformal modulus. Note that this, together with conformal invariance of the extremal distance, gives yet another proof of Theorem 2.49.

By conformal invariance of extremal distance and conformal modulus, it is sufficient to prove the identity for an annulus $A(r,R)$. We will treat the general case $0 < r < R < \infty$ and leave the cases $r = 0$ and/or $R = \infty$ to the reader.

As in the case of rectangles, it is easy to guess the extremal metric. Taking $\rho_0 = 1/|z|$ and considering curves along the radial directions, we have $L(\Gamma, \rho_0) = \int_r^R (1/t)\mathrm{d}t = \log(R/r)$ and $A(\rho_0) = \int_r^R \int_0^{2\pi} (1/t)\mathrm{d}t\mathrm{d}\theta = 2\pi \log(R/r)$. Hence

$$\lambda(\Gamma) \geq \frac{L(\Gamma, \rho_0)}{A(\rho_0)} = \frac{\log^2(R/r)}{2\pi \log(R/r)} = \frac{1}{2\pi} \log(R/r).$$

To show that this metric is extremal, we again use Theorem 4.4, with Γ_0 being the family of straight radial intervals connecting two boundary components.

The same extremal metric and essentially the same argument give that

$$\lambda(\Gamma^*) = \frac{2\pi}{\log(R/r)},$$

where Γ^* is the conjugated family of the curves that separate two boundary components. In particular, as in the case of rectangles, we have $\lambda(\Gamma)\lambda(\Gamma^*) = 1$.

Conformal radius. Let Ω be a simply connected domain and z_0 be a finite interior point. It might seem that extremal lengths cannot give any meaningful information about such configurations. Indeed, by the Riemann theorem, all configurations like this are conformally invariant. But we do have a notion of conformal radius, which is defined in terms of conformal maps and which gives important information about this configuration.

Let us recall that if $f : \mathbb{D} \to \Omega$ is a univalent map with $f(0) = z_0$, then the conformal radius $R = R(\Omega, z_0)$ of Ω with respect to z_0 is $|f'(0)|$. Alternatively, there is a univalent function $g : \Omega \to R\mathbb{D}$ such that its Taylor series at z_0 is of the form

$$g(z) = (z - z_0) + a_1(z - z_0)^2 + \cdots . \tag{4.2}$$

It turns out that it is possible to define the conformal radius in terms of the extremal length. For this, we have to introduce the notion of the *reduced modulus* of Ω with respect to z_0. The reduced modulus $m(\Omega, z_0)$ is defined as

$$\lim_{\varepsilon \to 0} \left(M(\Omega_\varepsilon) + \frac{1}{2\pi} \log(\varepsilon) \right),$$

where $M(\Omega_\varepsilon)$ is the conformal modulus of the doubly connected domain $\Omega_\varepsilon = \Omega \setminus B(z_0, \varepsilon)$ (this is well defined for ε sufficiently small).

It is possible to use comparison to show directly that this limit exists. Here, we will follow a different strategy which not only shows that the limit exists but also allows us to compute it.

Let g be as in (4.2). By conformal invariance, $M(\Omega_\varepsilon) = M(g(\Omega_\varepsilon))$. By (4.2),

$$\{\varepsilon - o(\varepsilon) < |z| < R\} \subset g(\Omega_\varepsilon) \subset \{\varepsilon + o(\varepsilon) < |z| < R\}.$$

This means that $g(\Omega_\varepsilon)$ is squeezed between two annuli. By monotonicity of extremal lengths,

$$\frac{1}{2\pi}\log\left(\frac{R}{\varepsilon + o(\varepsilon)}\right) \leq M(g(\Omega_\varepsilon)) \leq \frac{1}{2\pi}\log\left(\frac{R}{\varepsilon - o(\varepsilon)}\right).$$

Adding $(1/2\pi)\log(\varepsilon)$ to all parts, we have

$$\frac{1}{2\pi}\log\left(\frac{R}{1 + o(1)}\right) \leq M(g(\Omega_\varepsilon)) + \frac{1}{2\pi}\log(\varepsilon) \leq \frac{1}{2\pi}\log\left(\frac{R}{1 - o(1)}\right).$$

This allows one to pass to the limit and to show that $m(\Omega, z_0) = (1/2\pi)\log(R)$. This also shows that the conformal radius could be formulated in terms of the extremal length.

An interior point and a boundary arc. Let Ω be a simply connected domain, z_0 be a point inside, and A be a boundary arc. We can consider two families of curves Γ and Γ^*. The first family consists of curves that begin and end on A and go around z_0; the second family consists of all curves that separate A from z_0 (see Figure 4.4). Both $\lambda(\Gamma)$ and $\lambda(\Gamma^*)$ are conformal invariants of the configuration (Ω, z_0, A). But we know that the conformal type of such configuration is uniquely determined by the harmonic measure $\omega_\Omega(z_0, A)$. This proves that $\lambda(\Gamma)$ and $\lambda(\Gamma^*)$ can be written as functions of harmonic measure.

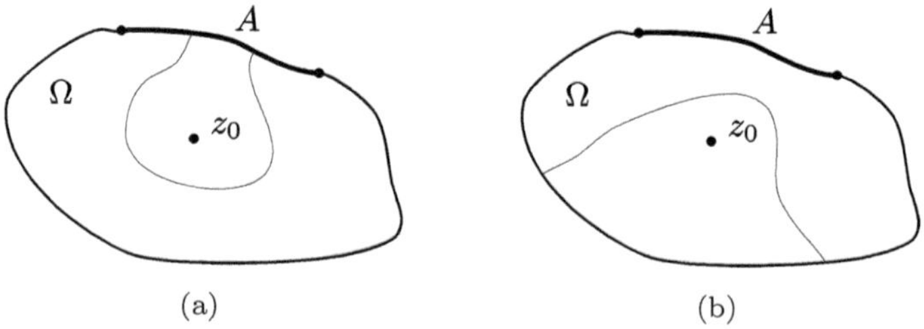

Fig. 4.4. (a) Γ is the family of all curves that start and end on A and go around z_0. (b) Γ^* is the dual family which separates z_0 from A.

We start by reducing this extremal length to a certain extremal distance. There is a univalent map $f : \Omega \to \mathbb{D}$ such that $f(z_0) = 0$ and the image of the arc is $E = \{e^{i\theta}, -\theta_0 \le \theta \le \theta_0\}$. This configuration and the corresponding family of curves Γ are symmetric with respect to the real line. By the symmetry rule, it is sufficient to consider only symmetric metrics.

Let $\gamma(t) = (x(t), y(t))$ be any curve from Γ. Without loss of generality, we can assume that its starting point is in the upper half-plane. Let us define $\gamma_+(t) = (x(t), |y(t)|)$ with $0 \le t \le t_+$, where t_+ is the first t such that $-1 < x(t) < 0$ and $y(t) = 0$, that is, the first time γ crosses the interval $[-1, 0]$. After this, we can join together γ_+ and its symmetric image to form a symmetric curve from Γ. In a similar way, we can construct $\gamma_-(t) = (x(t), -|y(t)|)$ with $t_- < t$ where t_- is the last time the curve γ intersects $[-1, 0]$. As before, we join γ_- with its symmetric image (see Figure 4.5). It is clear that for any symmetric metric, at least one of these symmetric curves has the length bounded by the length of γ. This proves that in the definition of $L(\Gamma, \rho)$ we can consider only symmetric curves that do not cross $[0, 1]$ (but may touch it) and cross $[-1, 0]$ only once.

By symmetry, the area of the disc is twice the area of the upper half-disc $\mathbb{D}_+$ and the length of a symmetric curve is twice the length of its upper half. This proves that $\lambda(\Gamma) = 2\lambda(\Gamma')$, where Γ' is the family of curves in the upper half-disc that connect the arc $E_+ = \{e^{i\theta}, 0 \le \theta \le \theta_0\}$ with the interval $F = [-1, 0]$. By definition, $\lambda(\Gamma') = d_{\mathbb{D}_+}(E_+, F)$.

The same argument with symmetry (or as shown in Exercise 4.3) shows that $\lambda(\Gamma) = 4d_\Omega(E, F)$, where $\Omega = \mathbb{D} \setminus [-1, 0]$.

Remark 4.9. As a byproduct of this argument, we obtain that

$$\lambda(\Gamma)\lambda(\Gamma^*) = 4,$$

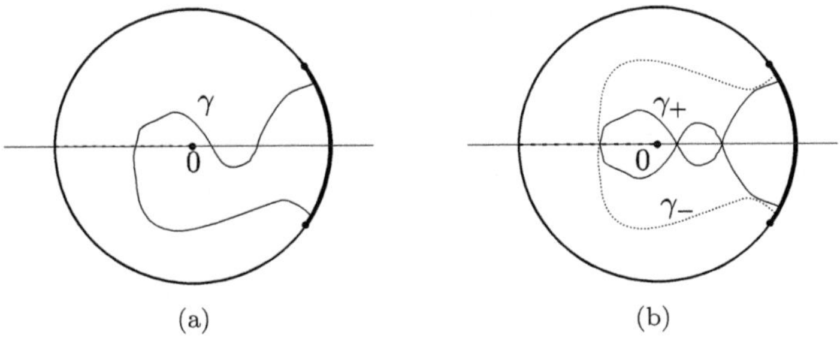

(a) (b)

Fig. 4.5. (a) A curve γ (b) and symmetrized curves γ_+ (solid) and γ_- (dotted).

where Γ and Γ^* are as in Figure 4.4. As we have seen above, $\lambda(\Gamma) = 2d_{\mathbb{D}_+}([-1,0], E_+)$, where $E_+ = \{e^{i\theta}, 0 \le \theta \le \theta_0\}$ and $\theta_0 = \pi\omega_\Omega(z_0, A)$. Since Γ^* is essentially the same family of curves for the complementary boundary arc $\tilde{A} = \partial\Omega \setminus A$, the same argument shows that $\lambda(\Gamma^*) = 2d_{\mathbb{D}_+}([0,1], \tilde{E}_+)$, where $\tilde{E}_+ = \{e^{i\theta}, \theta_0 \le \theta \le \pi\}$. Thus, $\lambda(\Gamma)$ and $\lambda(\Gamma^*)$ are expressed in terms of conjugated extremal distances whose product is always 1. This proves that $\lambda(\Gamma)\lambda(\Gamma^*) = 4$.

Exercise 4.5. Compute the extremal length of Γ in terms of θ_0 (note that $\omega_\Omega(z_0, A) = \theta_0/\pi$).

There is an alternative way to approach this problem. Following Beurling (1989, p. 369), we consider

$$\lambda(z_0, A) = \sup_\gamma d_\Omega(\gamma, A), \tag{4.3}$$

where the supremum is over all curves γ connecting z_0 with the boundary of Ω. This quantity is obviously a conformal invariant. The benefit of this quantity is that we take the supremum over γ and the extremal distance is the supremum over all metrics. This means that any choice of γ and metric ρ gives a lower bound on this quantity.

Lemma 4.10. *Let Ω be a simply connected domain, z_0 be a point inside and A be a boundary arc. Then $\lambda(z_0, A) = \lambda(\Gamma)/4$, where Γ is a family of curves that start and end on A and go around z_0.*

Proof. By conformal invariance, we can consider the same model case as before: $\Omega = \mathbb{D}$, $z_0 = 0$, and $A = \{e^{i\theta}, -\theta_0 \le \theta \le \theta_0\}$. Let γ be any simple curve connecting 0 with the complementary arc $\partial\mathbb{D} \setminus A$. Using this curve as a branch cut, we can define two branches of the square root in Ω. Two branches will map Ω onto two disjoint domains Ω_1 and $\Omega_2 = -\Omega_1$ such that the union of their closures is the closed unit disc. The curve γ will be mapped onto a symmetric curve γ' which separates Ω_1 and Ω_2 inside the unit disc. The arc A will be mapped to two symmetric arcs A_1 and A_2 (see Figure 4.6). By conformal invariance, $d_\Omega(\gamma, A) = d_{\Omega_i}(\gamma', A_i)$. By the serial rule (Proposition 4.6),

$$d_{\mathbb{D}}(A_1, A_2) \ge d_{\Omega_1}(\gamma', A_1) + d_{\Omega_2}(\gamma', A_2) = 2d_\Omega(\gamma, A).$$

Note that, independent of γ, the arcs A_1 and A_2 are $\{e^{i\theta}, |\theta| \le \theta_0/2\}$ and $\{e^{i\theta}, |\theta - \pi| \le \theta_0/2\}$. Hence, the left-hand side of the inequality is

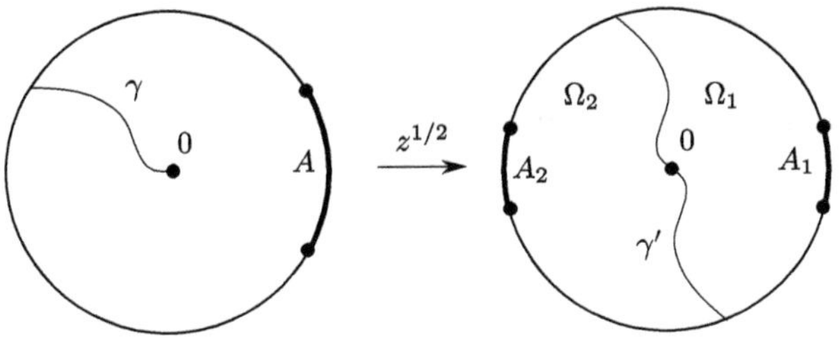

Fig. 4.6. Under multiply-valued $z^{1/2}$ the arc A is mapped to two arcs A_1 and A_2, each of which is half the length of A.

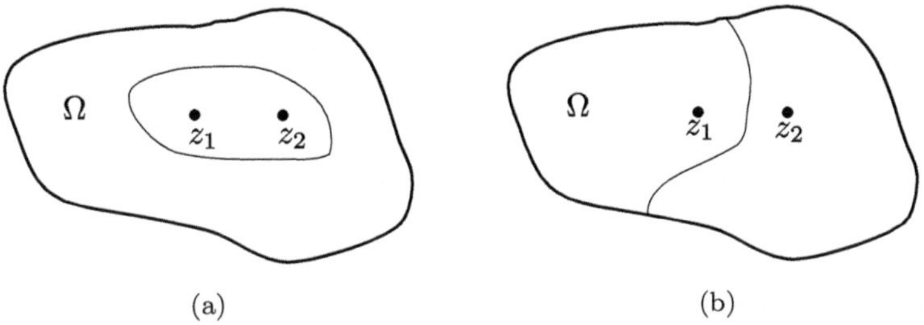

Fig. 4.7. (a) Γ is the family of loops that separate z_1 and z_2 from the boundary. (b) The dual family Γ^* is made of curves separating z_1 from z_2.

independent of γ and maximizing over γ, we obtain

$$\lambda(0, A) \leq \frac{1}{2} d_{\mathbb{D}}(A_1, A_2).$$

But when $\gamma = [-1, 0]$ and $\gamma' = [-i, i]$ we have an equality in the serial rule, and hence

$$\lambda(0, A) = d_{\mathbb{D}}([-1, 0], A).$$

By the symmetry argument identical to the one preceding this lemma, $\lambda(\Gamma) = 4d_{\mathbb{D}}([-1, 0], A)$. This completes the proof of the lemma. □

Two interior points. In the case of two interior points z_1 and z_2, we can consider two families Γ and Γ^* of loops surrounding two points and curves connecting boundary points and separating points z_1 and z_2 from each other (see Figure 4.7).

By conformal invariance, it is sufficient to consider the case when $\Omega = \mathbb{D}$, one of the marked points is 0 and the other one is $x \in (0, 1)$. Since our configuration is symmetric with respect to $z \mapsto \bar{z}$, we can apply the symmetry rule and consider only symmetric metrics. As before, using symmetry, one can show that it is enough to consider symmetric curves, and extremal lengths can be reduced to extremal distances.

The argument is very similar to the one in the preceding section; by using symmetry, for every loop going around two points we can find a loop which has the same length and goes around the interval $[0, x]$. Then we can construct a loop which is symmetric and has the same or smaller length.

We start with a curve separating 0 and x from the unit circle; see Figure 4.8(a). By flipping excursions with respect to the real line, we can construct a curve of the same length which goes once around the interval $[0, x]$ and crosses the real line only once. Such a curve, as in Figure 4.8(b), has upper and lower parts. We can consider two curves that are given by the union of the upper part with its symmetric image and the lower part with its symmetric image. One of these two curves will have length equal to or smaller than the length of the original curve. Thus, we can construct a shorter *symmetric* curve; see Figure 4.8(c).

Let $\tilde{\Gamma}$ be the family of curves separating $[0, x]$ and the unit circle. The argument above proves that for a symmetric metric ρ, for each curve γ separating 0 and x from the unit circle, there is a curve $\tilde{\gamma} \in \tilde{\Gamma}$ which is at most as long as γ. In other words, $L(\Gamma, \rho) \geq L(\tilde{\Gamma}, \rho)$. But since $\tilde{\Gamma} \subset \Gamma$, we have $L(\Gamma, \rho) \leq L(\tilde{\Gamma}, \rho)$. Since this is true for every symmetric ρ and we know that to compute the extremal length, it is sufficient to consider symmetric ρ, and we have $\lambda(\Gamma) = \lambda(\tilde{\Gamma})$.

As before, this extremal length is twice the modulus of a conformal rectangle which is the upper half-disc with vertices -1, 0, x, and 1 (extremal distance between $[-1, 0]$ and $[x, 1]$ inside the upper half-disc). By using the Joukowsky map, it can be mapped onto the upper half-plane with four marked points. Using essentially the same argument as in Exercise 4.5, we can compute this conformal modulus in terms of x.

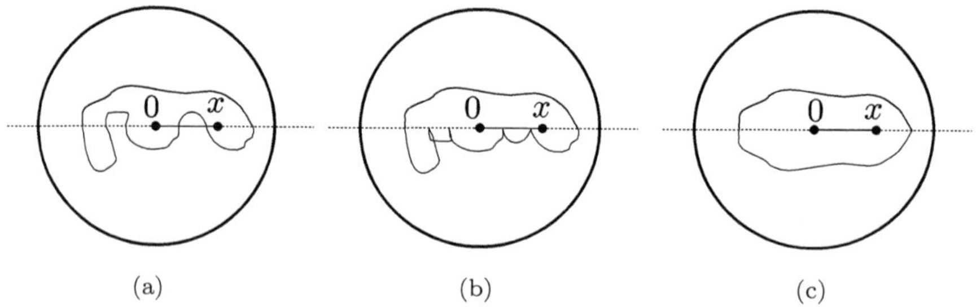

Fig. 4.8. Construction of a shorter symmetric curve separating the interval $[0, x]$ from the unit circle.

The extremal length $\lambda(\Gamma)$ defines the conformal type of a domain with two marked points. But Green's function also defines this type. This means that $G_\Omega(z_1, z_2)$ must be a function of $\lambda(\Gamma)$. This relation can be written explicitly since we know how to write $\lambda(\Gamma)$ in terms of x and $x = \exp(-G_\Omega(z_1, z_2))$.

Exercise 4.6. Use extremal lengths to show that the Green function is symmetric, namely, that $G_\Omega(z_1, z_2) = G_\Omega(z_2, z_1)$.

Exercise 4.7. Let Γ and Γ^* be as in Figure 4.7. Show that $\lambda(\Gamma)\lambda(\Gamma^*) = 4$.

Extremal domains. The argument above allows us to connect the extremal length $\lambda(\Gamma) = \lambda(\tilde{\Gamma})$ with certain extremal problems.

Let us consider a doubly connected domain $\Omega = \mathbb{D} \setminus [0, x]$. Then $\lambda(\tilde{\Gamma})$ is the inverse of its modulus.

We also note that if s is any set connecting 0 and x inside $\mathbb{D}$ such that $\Omega_s = \mathbb{D} \setminus s$ is a doubly connected domain and Γ_s is a family of curves separating s from the unit circle, then $\Gamma_s \subset \Gamma$ and $\lambda(\Gamma) \leq \lambda(\Gamma_s)$. Comparing with the previous identity, we have that $\lambda(\Gamma)$ is the minimum of $\lambda(\Gamma_s)$ and the minimum is achieved for $s = [0, x]$. Equivalently, out of all doubly connected domains separating 0 and x from the unit circle, the domain $\Omega = \mathbb{D} \setminus [0, x]$ has the largest modulus. In a sense, this is a very natural statement; $[0, x]$ is the set connecting 0 and x which stays as far from $\partial\mathbb{D}$ as possible.

Applying the map $1/z$, conformal invariance and noticing that the modulus of a doubly connected domain is the same thing as the extremal distance between its boundaries, we immediately have the following theorem.

Theorem 4.11 (Grötzsch). *Of all continua that connect the point $R > 1$ with infinity, the ray $[R, +\infty]$ has the greatest extremal distance from the unit circle.*

The domain $\mathbb{D}_- \setminus [R, +\infty]$ is called the Grötzsch annulus.

There is another extremal problem very similar to the Grötzsch theorem: Given three points, what is the largest doubly connected domain separating two of them from the third one and infinity? This problem was solved by Teichmüller (1938). The precise statement is given in the following theorem.

Theorem 4.12 (Teichmüller). *Let Ω be a doubly connected domain and E_1 and E_2 be components of its complement. We assume that -1 and 0 belong to the bounded component E_1 and that the unbounded component E_2 contains a point w_0 with $|w_0| = R$. Then the modulus of Ω is bounded by the modulus of $\Omega_R = \mathbb{C} \setminus ([-1, 0] \cup [R, +\infty])$.*

The extremal domain Ω_R is called the Teichmüller annulus. Theorem 4.12 could be reformulated as the statement that out of all doubly connected domains that separate $\{-1, 0\}$ from infinity and some point with modulus R, the Teichmüller annulus has the largest modulus.

Proof of Theorem 4.12. Let us consider a simply connected domain $\widetilde{\Omega} = \Omega \cup E_1$. There is a univalent map $f : \mathbb{D} \to \widetilde{\Omega}$ with $f(0) = 0$.

Let z_0 be the point mapped to -1. By the Growth theorem (Theorem 3.11) applied to the rescaled f, we have

$$1 = |f(z_0)| \leq \frac{|z_0||f'(0)|}{(1 - |z_0|)^2} \leq \frac{4R|z_0|}{(1 - |z_0|)^2}, \tag{4.4}$$

where the last inequality follows from the Koebe 1/4 theorem. By this theorem, the image of $\mathbb{D}$ contains a disc of radius $|f'(0)|/4$, hence $R = |w_0| \geq |f'(0)|/4$.

In the case of the Teichmüller annulus, $E_1 = [-1, 0]$ and $E_2 = [R, \infty]$. It is easy to check that the function $g(z) = -4RK(z)$, where K is the Koebe function, maps the unit disc onto the complement of E_2. The preimage of -1 under this map is a negative real number which we denote by z_R. Both Growth and Koebe 1/4 theorems are sharp in this case; hence in (4.4), we have equalities instead of both inequalities. Alternatively, since the Koebe function is known explicitly, the identity can be easily verified algebraically. Combining this equality with (4.4), we get

$$\frac{4R|z_R|}{(1 - |z_R|)^2} = 1 \leq \frac{4R|z_0|}{(1 - |z_0|)^2}.$$

Since $x/(1 - x)^2$ is an increasing function on $[0, 1]$, we obtain $|z_R| \leq |z_0|$.

By conformal invariance, the modulus of Ω is equal to the extremal distance between $F_1 = f^{-1}(E_1)$ and the unit circle. The set F_1 connects z_0 and 0. By Theorem 4.11 or the argument preceding it, for given $|z_0|$, the maximum modulus is achieved for $F_1 = [0, z_0]$. Since the modulus of the domain $\mathbb{D} \setminus [0, |z_0|]$ is monotone in $|z_0|$ and $|z_0| \geq |z_R|$, we prove that the modulus of Ω is bounded by the modulus of Ω_R. Hence, indeed, the modulus is maximized by the Teichmüller annulus. □

The Teichmüller theorem has a simple geometrical corollary which we leave as an exercise.

Exercise 4.8. Let Ω be a doubly connected domain such that its modulus is strictly greater than 1/2. Show that there is a circle inside Ω which separates the boundary components.

4.2.2. *Geometric application*

The exercise above is a good example of how one can derive interesting geometric conclusions from conformal invariants. Here, we give two more examples of how one can use extremal length to obtain purely geometrical inequalities that are extremely difficult to obtain by other means.

Conformal rectangles. Let Ω be a conformal rectangle and denote the four boundary arcs by E, E', F, F'. Then

$$d(E,F)d(E',F') \le A(\Omega),$$

where d is the Euclidean distance between sets and A is the area of Ω.

The proof of this inequality is almost trivial. Let Γ be the family of curves connecting E and F, and Γ^* be the family of curves connecting E' and F'. Let us consider $\rho = 1$ in Ω, that is, the usual Euclidean metric in Ω. For this metric, $L(\Gamma,\rho) \ge d(E,F)$ and $L(\Gamma^*,\rho) \ge d(E',F')$; hence $\lambda(\Gamma) \ge L^2(\Gamma,\rho)/A(\Omega) \ge d^2(E,F)/A(\Omega)$ and the similar inequality holds for $\lambda(\Gamma^*)$. For conformal rectangles, we have that $\lambda(\Gamma)\lambda(\Gamma^*) = 1$, which together with the previous inequalities implies $d^2(E,F)d^2(E',F') \le A^2(\Omega)$.

Conformal triangles. Let Ω be a conformal triangle, that is, a simply connected domain with three marked points. Then there is a curve γ in Ω such that γ touches all three sides of Ω and

$$\text{length}(\gamma) \le \sqrt[4]{3}\sqrt{A(\Omega)}.$$

Moreover, the constant $\sqrt[4]{3}$ is sharp.

Let Γ be the family of all curves that touch all three sides of Ω. We would like to compute the extremal length of this family. The case of conformal triangles is different from all previous examples that we have considered so far: all conformal triangles are conformally equivalent. This means that $\lambda(\Gamma)$ is just an absolute constant which turns out to be $\sqrt{3}$ (see Exercise 4.9). Now, let Ω be an arbitrary conformal triangle and let us consider $\rho = 1$ in Ω. Then

$$\sqrt{3} = \lambda(\Gamma) \ge \frac{\min(\text{length}^2(\gamma))}{A(\Omega)}.$$

This is equivalent to the desired inequality. To see that the constant is sharp, we consider the equilateral triangle as above and γ is one of the triangle's altitudes.

Exercise 4.9. Prove that the extremal length of the family of curves connecting three sides of a conformal triangle is $\sqrt{3}$.

Exercise 4.10. Let Ω be a conformal triangle. Denote its vertices by a, b, and c and 'sides' by (a, b), (b, c), and (c, d). By Γ, we denote the family of curves in Ω that have endpoints on (a, c) and (b, c) and pass through (a, b). Show that $\lambda(\Gamma) = 2$.

Exercise 4.11. Let z_1 and z_2 be two distinct points in $\mathbb{C}$ and let Γ be the family of all rectifiable curves that form figure eights around z_1 and z_2, namely, all oriented curves that have winding number 1 with respect to one of the points and -1 with respect to another. Note that this property is independent of orientation. Show that $\lambda(\Gamma) = 4$.

4.3. Harmonic Measure Revisited

We have already seen that there is a connection between the harmonic measure and the extremal length. In this section, we continue to investigate their relationship. We start with the simplest case of a rectangle:

$$R = \{z : -L < \operatorname{Re} z < L, \ -1 < \operatorname{Im} z < 1\},$$

and we would like to compute the harmonic measure of its vertical sides at the center of the rectangle.

We have an explicit conformal map from $\mathbb{H}$ onto R, and, composing it with a Möbius transformation, we obtain a map from $\mathbb{D}$ to R. The harmonic measure on $\mathbb{D}$ is just the normalized arc length. Using this argument, we should be able to find the harmonic measure explicitly. Unfortunately, this computation involves very unpleasant manipulations with elliptic functions and in reality is not very fruitful. Instead, we write a very good *estimate* of harmonic measure. This estimate is well known and has appeared in many papers. Here, we follow a very nice presentation from Garnett and Marshall (2008, IV.5).

Lemma 4.13. *Let R be a rectangle as above and let $E = \{z : \operatorname{Re} z = -L,$ $-1 < \operatorname{Im} z < 1\}$ be its left side. Then*

$$\frac{1}{2}e^{-\pi L/2} \leq \omega(0, E) \leq \frac{4}{\pi}e^{-\pi L/2}.$$

Moreover, the first inequality is sharp in the limit $L \to 0$ and the second one is sharp in the limit $L \to \infty$.

Remark 4.14. A simpler fact that $\omega(0, E) \asymp e^{-\pi L/2}$ as $L \to \infty$ could be easily extracted from the Christoffel–Schwarz formula.

Proof of Lemma 4.13. Instead of R, we consider a much simpler domain $S = \{z : -L < \operatorname{Re} z,\ -1 < \operatorname{Im} z < 1\}$. Let us consider the map

$$f(z) = \sin\left(\frac{i\pi}{2}(z+L)\right)$$

which maps S onto $\mathbb{H}$ and E onto $[-1,1]$. We already know that the density of the harmonic measure $\omega_{\mathbb{H}}(f(z), \mathrm{d}t)$ is given by the Poisson kernel. By symmetry of the Poisson kernel and the elementary properties of sine for each vertical interval in S, the maximum harmonic measure is attained at its centre. In particular, on $[-i,i]$, the maximum harmonic measure is attained at $z=0$. Clearly, $f(0) = i\sinh(L\pi/2)$ and the explicit integration of the Poisson kernel gives us that

$$\omega_S(0,E) = \frac{2}{\pi}\arctan\left(\frac{1}{\sinh(L\pi/2)}\right) = \frac{4}{\pi}\arctan(e^{-\pi L/2}), \tag{4.5}$$

where the last equality follows from the double angle formula.

By monotonicity of harmonic measure with respect to the domain, we have $\omega_R(z,E) \le \omega_S(z,E)$ and, in particular, $\omega_R(0,E) \le \omega_S(0,E)$. But on the boundary of R, we have

$$\omega_S(z,E) + \omega_S(-z,E) \le (\omega_R(z,E) + \omega_R(-z,E)) \times \left(1 + \sup_{z\in[L-i,L+i]} \omega_S(z,E)\right)$$

and by the maximum principle the same is true inside, in particular, at $z = 0$. We also note that by essentially the same argument as the one preceding (4.5),

$$\sup_{z\in[L-i,L+i]} \omega_S(z,E) = \omega_S(L,E) = \frac{4}{\pi}\arctan(e^{-\pi L}). \tag{4.6}$$

Combining all the estimates together, we obtain

$$\frac{\omega_S(0,E)}{1+\omega_S(L,E)} \le \omega_R(0,E) \le \omega_S(0,E).$$

Plugging in (4.5) and (4.6), we rewrite our estimates as

$$\frac{\frac{4}{\pi}\arctan(e^{-\pi L/2})}{1+\frac{4}{\pi}\arctan(e^{-\pi L})} \le \omega_R(0,E) \le \frac{4}{\pi}\arctan(e^{-\pi L/2}). \tag{4.7}$$

For $0 < x < 1$, we can estimate $\pi x/4 \le \arctan(x) \le x$ and $\arctan(x) \le \pi/4$. Using these estimates, we get

$$\frac{1}{2}e^{-\pi L/2} \le \omega_R(0,E) \le \frac{4}{\pi}e^{-\pi L/2}.$$

Since $e^{\pi L}\arctan(e^{-\pi L}) \to 1$ for $L \to \infty$, we see from (4.7) that

$$\lim_{L\to\infty} e^{\pi L}\omega_R(0,E) = \frac{4}{\pi},$$

which means that the upper bound is sharp in this limit.

To see that the lower bound is sharp, we observe that by symmetry and scaling invariance $\omega_R(0,E) = 1/2 - \omega_{R'}(0,E')$, where $\mathbb{R}'$ is the similar rectangle with L replaced by $1/L$ and E' is its left side. □

Since L is also the extremal distance between the vertical sides of the rectangle, this lemma gives an asymptotic relation between the harmonic measure and the extremal distance. We can use Lemma 4.10 to generalize it to the case of arbitrary domains.

Theorem 4.15. *Let Ω be a simply connected domain, z_0 be a point inside and A be a boundary arc. Then*

$$e^{-\pi L/2} \le \omega_\Omega(z_0, A) \le \frac{8}{\pi}e^{-\pi L/2},$$

where $L = \lambda(\Gamma)/2 = 2\lambda(z_0, A)$. Here, Γ is the family of curves in Ω that start and end on A and go around z_0 and $\lambda(z_0, A)$ is the conformal invariant defined by (4.3).

Proof. The proof is just a combination of Lemmas 4.10 and 4.13. As in the proof of Lemma 4.10, we construct a one-to-two map from Ω onto a rectangle R such that A goes to two vertical sides and z_0 goes to the centre of R, which we assume to be 0. The side length ratio L is equal to $\lambda(\Gamma)/2 = 2\lambda(z_0, A)$.

But by conformal invariance, $\omega_\Omega(z_0, A) = 2\omega_R(0, E)$, where E is one of the vertical sides. Combining this with Lemma 4.13, we complete the proof of the theorem. □

This theorem gives us a way to estimate the harmonic measure in terms of the extremal length, but the computation of the extremal length is a non-trivial problem. Theorem 4.16 gives a lower bound in terms of a more geometrical quantity.

Theorem 4.16. *Let $\phi(x) < \psi(x)$ be two continuous functions on $[a, b]$ and define a strip domain $\Omega = \{(x, y) : \phi(x) < y < \psi(x),\ a < x < b\}$. Then*

$$d_\Omega(E, F) \ge \int_a^b \frac{1}{\theta(x)}\mathrm{d}x,$$

where E and F are the left and the right vertical sides of Ω and $\theta(x) = \psi(x) - \phi(x)$ is the width of the strip at x.

The estimate is sharp if ϕ and ψ are constant functions. Note that θ being constant is not sufficient.

Proof. Let us define a metric $\rho_0(x,y) = 1/\theta(x)$. Each curve connecting E and F must cross each vertical section of Ω. Hence we have a lower bound

$$L(\Gamma, \rho_0) \geq \int_a^b \frac{1}{\theta(x)} \mathrm{d}x.$$

The area can be computed explicitly:

$$A(\Omega, \rho_0) = \int_a^b \int_{\phi(x)}^{\psi(x)} \frac{1}{\theta^2(x)} \mathrm{d}y \mathrm{d}x = \int_a^b \frac{1}{\theta(x)} \mathrm{d}x.$$

Together with the previous estimate, this gives us the required result. □

The proof of this theorem is completely elementary, but it looks like a trick which is not very illuminating. To explain the main idea behind the theorem, we slice the strip Ω into many vertical, almost rectangular, domains $\Delta(x) \times [\phi(x), \psi(x)]$. The extremal distance between the vertical sides is approximately $\Delta(x)/\theta(x)$. By the serial rule, the extremal distance between E and F should be at least

$$\sum \frac{\Delta(x)}{\theta(x)} \approx \int_a^b \frac{1}{\theta(x)} \mathrm{d}x.$$

This motivates the statement of the theorem. (The idea of using this integral is a bit older than the theorem itself and goes back to Carleman.)

To explain the choice of ρ_0, we note that every constant metric is an extremal metric in a rectangle. This suggests that we should consider a metric which depends on x only. With metric ρ_0, each infinitesimal rectangle looks like rectangle with height 1, and so they glue together naturally. This gives a hand-waving explanation why we consider this particular metric ρ_0.

We also would like to note that the same argument is valid for any strip domain: we just have to define θ to be the distance between the highest point on the lower boundary and the lowest point on the top boundary. Another observation is that we do not really need continuity: we just need θ to be measurable.

One of the main corollaries is obtained by considering a polar version of this estimate and combining it with harmonic measure estimates.

Theorem 4.17 (Ahlfors). *Let Ω be a simply connected domain and let ζ be an accessible point on the boundary which is defined by a rectifiable* *curve γ which connects some reference point z_0 inside Ω with ζ. Then*

$$\omega(z_0, U(\zeta, r_0)) \leq \frac{8}{\pi} \exp\left(-\pi \int_{r_0}^{R} \frac{dr}{r\theta(r)}\right), \tag{4.8}$$

where $U(\zeta, r)$ is the component of $B(\zeta, r) \cap \Omega$ which contains the tail of γ and $\theta(r)$ is the angular size of the arc A_r of $\{|\zeta| = r\} \cap \Omega$ which intersects the tail of γ (note that $r\theta(t)$ is the length of this arc).

Proof. Without loss of generality, we assume that Ω is bounded. Since ζ is on the boundary of a simply connected domain Ω, we can define a single valued branch of $\log(z - \zeta)$ in Ω. The image domain Ω' could be considered as a semi-infinite strip domain. By conformal invariance and monotonicity of the harmonic measure,

$$\omega(z_0, U(\zeta, r_0)) \leq \omega_{\Omega'_r}(w_0, I_{r_0}),$$

where Ω'_r is the intersection of Ω' with $\{\operatorname{Re} z > \log(r_0)\}$, $w_0 = \log(z_0 - \zeta)$, and I_r is a vertical interval which is the image of A_r.

By Theorem 4.15, we have

$$\omega_{\Omega'_r}(w_0, I_{r_0}) \leq \frac{8}{\pi} e^{-\pi\lambda},$$

where $\lambda = \lambda(w_0, I_{r_0})$. Consider a cut σ from w_0 which lies completely to the right of w_0. For this choice of a cut, every curve connecting the cut with I_{r_0} also connects I_{r_0} with I_R, where $R = |z_0 - \zeta|$. By monotonicity of the extremal length,

$$\lambda \geq d(\sigma, I_{r_0}) \geq d(I_R, I_{r_0}).$$

By Theorem 4.16,

$$d(I_R, I_{r_0}) \geq \int_{\log(r_0)}^{\log(R)} \frac{dt}{\theta(e^t)}.$$

Combining together all estimates, we obtain

$$\begin{aligned}\omega(z_0, U(\zeta, r_0)) &\leq \frac{8}{\pi} \exp\left(-\pi \int_{\log(r_0)}^{\log(R)} \frac{dt}{\theta(e^t)}\right) \\ &= \frac{8}{\pi} \exp\left(-\pi \int_{r_0}^{R} \frac{dr}{r\theta(r)}\right).\end{aligned}$$

□

Simple estimates of $\theta(r)$ give important estimates of the scaling behaviour of the harmonic measure. This result is known as the Beurling estimate and provides a simple, universal estimate of the harmonic measure of a set in terms of set's diameter. Here, we derive it as a corollary of Theorem 4.17; in many books, it is a corollary of a result known as the *Beurling projection theorem.* Interested readers can find the details in Garnett and Marshall (2008, Section III.9).

Corollary 4.18. *Let Ω, ζ, and $U(\zeta,r)$ be as in the previous theorem. Then*

$$\omega(z_0, U(\zeta,r)) \leq \frac{8}{\pi\sqrt{R}}\sqrt{r},$$

where $R = |z_0 - \zeta|$.

This follows immediately from a trivial estimate $\theta(r) \leq 2\pi$. This corollary means that the harmonic measure of an r-neighbourhood of a boundary point decreases at least as the square root of the radius.

Exercise 4.12. Let $\Omega = \mathbb{H} \setminus [0,i]$, $z_0 = 2i$, and $\zeta = i$. Show that the harmonic measure from z_0 of an r-neighbourhood of ζ decays as $\sqrt{r}$. This proves that the previous estimate of the decay rate is sharp.

Corollary 4.19. *Let Ω be a simply connected domain. Let us consider two accessible points ζ_1 and ζ_2 that correspond to the same boundary point. Then*

$$\omega(z_0, U(\zeta_1,r))\omega(z_0, U(\zeta_2,r)) \leq \frac{8^2}{\pi^2 R^2}r^2.$$

This statement is also known as the Beurling estimate.

Proof of Corollary 4.18. We estimate both harmonic measures using Theorem 4.17:

$$\begin{aligned}
&\omega(z_0, U(\zeta_1,r))\omega(z_0, U(\zeta_2,r)) \\
&\quad \leq \frac{8^2}{\pi^2}\exp\left(-\pi\int_r^R \frac{1}{r}\left(\frac{1}{\theta_1(r)} + \frac{1}{\theta_2(r)}\right) \mathrm{d}r\right).
\end{aligned}$$

By harmonic mean–arithmetic mean inequality,

$$\frac{1}{\theta_1(r)} + \frac{1}{\theta_2(r)} \geq \frac{4}{\theta_1(r)+\theta_2(r)} \geq \frac{2}{\pi}.$$

Plugging this into the estimate of the product of harmonic measures gives us the desired estimate. □

Note that the similar statement holds for the product of several harmonic measures; the only thing that changes is the numerical constant.

This corollary tells us that two harmonic measures cannot be simultaneously large. If ζ_1 and ζ_2 are on two sides of an analytic arc, then the harmonic measures are absolutely continuous with respect to the arc length, and hence $\omega(z_0, U(\zeta_i, r)) \asymp r$ and the product is of order r^2.

There is an alternative way to approach these two corollaries. For a Borel measure μ, we can define its local dimension at z by

$$\dim_z \mu = \lim_{r \to 0} \frac{\log \mu(B(z,r))}{\log r},$$

whenever the limit exists. If the limit does not exist, then we can consider the upper and lower dimensions $\overline{\dim}_z$ and $\underline{\dim}_z$ that are given by lim sup and lim inf.

In this language, Corollary 4.18 means that $\underline{\dim}_\zeta \omega \geq 1/2$ and Corollary 4.19 means that $\overline{\dim}_{\zeta_1} + \overline{\dim}_{\zeta_2} \geq 2$.

Exercise 4.13. Show that for all $d \geq 1/2$, there are Ω and $\zeta \in \partial\Omega$ such that $\dim_\zeta \omega = d$.

This type of local analysis of harmonic measure is closely related to the study of integral means and coefficient problems mentioned in Section 3.7. Further information can be found in Makarov (1998).

Beurling's estimates give an upper bound on the harmonic measure in terms of the Euclidean size of the set. It is not difficult to show that there is no universal lower bound: a large boundary set which is inside a deep and narrow fjord might have arbitrarily small harmonic measure. A simple example is given in Figure 4.9. By monotonicity of harmonic measure, $\omega_{\Omega_\varepsilon}(0, A) \leq \omega_{\Omega_0}(0, B)$. Since γ_0 is a circular arc, the harmonic measure of B can be computed explicitly. Alternatively, it is easy to argue that, since the boundary of Ω_0 is piecewise smooth, the harmonic measure of B is of order ε. Either way, we can see that $\omega_{\Omega_\varepsilon}(0, A) \to 0$ as $\varepsilon \to 0$. This shows that there is no lower bound on harmonic measure in terms of the size of the set.

Although the example above shows that there are no universal lower bounds, it also explains the mechanism: the harmonic measure of a big set might be small if it is shielded by something else. If we ignore this scenario, then it is possible to obtain a lower bound. There are several ways of doing this. The simplest is presented in the following proposition, which is an adaptation of the argument from Garnett and Marshall (2008, p. 118); see also Garnett (1986, Lemma 6.1).

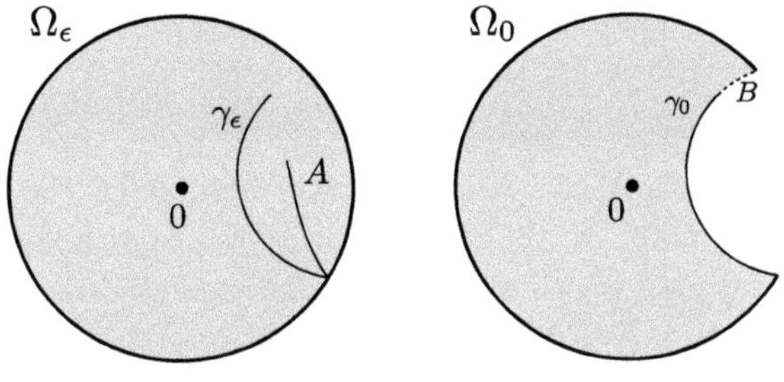

Fig. 4.9. Domain Ω_ε is the complement in $\mathbb{D}$ of a fixed cut A and a circular arc γ_ε which comes ε-close to the unit circle. Domain Ω_0 is obtained from $\mathbb{D}$ by cutting off the part which is separated from 0 by a circular arc γ_0. The set $B = \gamma_0 \setminus \gamma_\varepsilon$ is the dotted part of γ_0.

Proposition 4.20. *Let K be a connected closed set in the closed unit disc which intersects the unit circle; its complement is simply connected. Let $\zeta \in K \cap \partial\mathbb{D}$ and let r be the maximal value of $|z - \zeta|$ over all $z \in K$. If $r < r_0 < 1$, then $\omega_{\mathbb{D}\setminus K}(0, K) \geq cr$, where $c = c(r_0)$ is a constant which depends on r_0 but not on r.*

Explicit estimates of $c = c(r_0)$ can be found in the references above. For our purposes, it is sufficient to know that the harmonic measure is of order r if r is sufficiently small and the estimates are uniform if we assume that r is bounded away from 1.

Proof. In the proof, we use c to denote various constants, which are not necessarily the same but are independent of r. Let us consider a point $z_0 \in K$ such that $|z_0 - \zeta| = r$. Using rotation invariance, we can assume that $\arg(z_0) = 0$. Then either $1 - z_0 \geq cr$ or $|\arg(\zeta)| \geq cr$ for some universal constant c.

Let $\bar{K}$ be the complex conjugate of K, which is the set symmetric to K with respect to the radius passing through z_0. By Ω, we denote the component of $\mathbb{D} \setminus (K \cup \bar{K})$ which contains the origin. Applying twice the maximum principle for harmonic functions, we obtain

$$\omega_{\mathbb{D}\setminus K}(z, K) + \omega_{\mathbb{D}\setminus \bar{K}}(z, \bar{K}) \geq \omega_\Omega(z, K \cup \bar{K}) \geq \omega_{\mathbb{D}\setminus[z_0,1]}(z, A),$$

where A is the union of the interval $[z_0, 1]$ and the arc of the unit circle centred at 1 with endpoints ζ and $\bar{\zeta}$. By symmetry, if we plug in $z = 0$, we arrive at

$$2\omega_{\mathbb{D}\setminus K}(0, K) \geq \omega_{\mathbb{D}\setminus[z_0,1]}(0, A).$$

It is easy to show that if $1 - z_0 \geq cr$, then the harmonic measure of $[z_0, 1]$ is at least cr with some explicit constant c. Otherwise, $|\arg(\zeta)| \geq cr$ and in this case the harmonic measure of A is at least the harmonic measure of the arc, which is at least cr. Either way, we get

$$\omega_{\mathbb{D}\setminus[z_0,1]}(0, A) \geq cr,$$

which, together with the estimate above, completes the proof of the proposition. Constants can be made explicit by using the explicit formula for the conformal map which removes the slit (see (2.19)). □

Remark 4.21. Either composing with an appropriate Möbius transformation or by using essentially the same argument, one can show that if K is a half-plane hull of radius $r < r_0$ and $z_0 \in \mathbb{H}$ is such that $|z_0| \geq 10r_0$ and $\arg(z_0)$ is bounded away from 0 and π by at least $\varepsilon > 0$, then there is a universal constant $c = c(r_0, \varepsilon)$ such that

$$\omega_{\mathbb{H}\setminus K}(z_0, K) \geq c\frac{r}{\operatorname{Im} z_0}.$$

This lower bound on harmonic measure, together with an upper bound given by the Beurling estimate, implies an important result about the continuity of conformal maps onto nice domains. Roughly speaking, it gives a *uniform* bound on the distortion near the boundary. We state it in the disc case, but essentially the same is true for the half-plane hulls. The details are left to the reader.

Theorem 4.22. *Let $K_1 \subset K_2$ be two closed sets in the closure of a simply connected domain Ω such that their complements are simply connected. Let $z_0 \in \Omega \setminus K_2$. We assume that $B(z_0, r_0) \subset (\Omega \setminus K_2)$, and that the diameter of $K_2 \setminus K_1$ is less than $\varepsilon > 0$, where $\varepsilon < r_0/10$. Let g be a univalent map from $\Omega \setminus K_1$ onto $\mathbb{D}$ such that $g(z_0) = 0$ and let $A = g(K_2 \setminus K_1)$. Then there is a constant c which depends on r_0 only, such that the diameter of A is at most $c\sqrt{\varepsilon}$.*

Proof. Let ζ be an accessible point in $K_2 \setminus K_1$. Then by the maximum principle and by the Beurling estimate (Corollary 4.18),

$$\omega_{\Omega\setminus K_2}(z_0, K_2 \setminus K_1) \leq \omega_{\Omega\setminus K_2}(z_0, U(\zeta, 2\varepsilon)) \leq c\sqrt{\varepsilon},$$

where c is a constant depending on r_0 but not on ε. By conformal invariance of the harmonic measure,

$$\omega_{\mathbb{D}\setminus A}(0, A) \leq c\sqrt{\varepsilon}.$$

By Theorem 3.7, A is bounded away from 0 by some absolute constant (it depends only on the fact that ε is at least 10 times smaller than r_0). This allows the application of Proposition 4.20 to show that the diameter of A is at most $c\sqrt{\varepsilon}$, where c is another constant depending on r_0. □

4.4. Prime Ends Revisited

In this section, we give an alternative definition of prime ends in terms of extremal lengths. Conformal invariance of the extremal length will immediately give that the prime ends are also conformally invariant. This section follows the exposition of Ahlfors (2010). As before, for simplicity, we will assume that all domains are bounded, but this assumption is purely technical and can be dropped by using the spherical metric instead of the Euclidean one.

We will need a notion which is more general than a cross-cut. A *generalized cross-cut* or a *cluster of cross-cuts* is a finite collection of cross-cuts such that their union is connected.

Let Ω be a simply connected bounded domain and z_0 be some reference point inside. For a sequence (a_i) in Ω, we define Γ_a to be the collection of all cross-cut clusters that separate z_0 from all but finitely many a_i. As before, we will abuse notation and will call elements of Γ 'curves'. We say that a sequence (a_n) is *fundamental* if $\lambda(\Gamma_a) = 0$.

Although we define fundamental sequences in terms of some reference point z_0, the notion is independent of the choice of z_0. Let z_0' be another reference point and denote the corresponding collection of clusters by Γ_a'. We can connect z_0 and z_0' to be a continuous curve c inside Ω. There is $\delta > 0$ such that the distance between c and $\partial\Omega$ is $\delta > 0$.

Next, we note that $\lambda(\Gamma) = 0$ if and only if $L(\Gamma, \rho) = 0$ for every admissible metric ρ. This means that Γ contains arbitrarily short curves with respect to any metric with finite area. Let ρ_0 be a metric with finite area. Since Ω is bounded, $\rho = \rho_0 + 1$ is also an admissible metric. This means that for every $\varepsilon > 0$, there is $\gamma \in \Gamma_a$ such that its length with respect to ρ is at most ε. Hence, its length with respect to ρ_0 and 1 is also bounded by ε. Assuming that $\varepsilon < \delta$, we have that γ cannot intersect c, which implies that it separates a_n from z_0 if and only if it separates a_n from z_0'. This implies that $L(\Gamma_a', \rho_0) = 0$. Since this is true for every admissible ρ_0, we have that $\lambda(\Gamma_a')$.

We say that two fundamental sequences a and b are equivalent if their union is fundamental. Since the notion of the fundamental sequence is defined in terms of separating all but finitely many elements, the ordering of the sequence is not important and considering the union makes sense. Sometimes, to avoid ambiguities, it is easier to assume that the union of (a_n) and (b_n) is the sequence $a_1, b_1, a_2, b_2, \ldots$.

We have to show that this is indeed an equivalence relation. It is obvious that this relation is symmetric and reflexive. We only have to show that it is transitive. Let us assume that $(a_n) \sim (b_n)$ and $(b_n) \sim (c_n)$, which means that $\lambda(\Gamma_{a \cup b}) = \lambda(\Gamma_{b \cup c}) = 0$.

Let $\gamma_{a,b} \in \Gamma_{a \cup b}$ and $\gamma_{b,c} \in \Gamma_{b \cup c}$. If these two clusters intersect, then $\gamma_{a,b} \cup \gamma_{b,c}$ is a cluster which separates almost all elements of a, b, and c from z_0, which means that $\gamma_{a,b} \cup \gamma_{b,c} \in \Gamma_{a,c}$. This part of the argument is the reason why we consider cross-cut clusters instead of cross-cuts.

Assume that they do not intersect and consider the component of the complement of $\gamma_{a,b}$ which contains the reference point z_0. Denote this component by $\Omega_{a,b}$ and the corresponding component for $\gamma_{b,c}$ by $\Omega_{b,c}$. We consider that these components are 'outside' of the corresponding clusters and other components are 'inside'. Since the cross-cut clusters $\gamma_{a,b}$ and $\gamma_{b,c}$ do not intersect, we have three possibilities: each cluster is outside of the other one, $\gamma_{a,b}$ is inside of $\gamma_{b,c}$, or $\gamma_{b,c}$ is inside of $\gamma_{a,b}$.

Since almost all elements of (b_n) are inside both clusters, the first case cannot happen. In the second case, almost all elements of (a_n) are inside $\gamma_{a,b}$ which is inside of $\gamma_{b,c}$. This proves that almost all elements of all three sequences are separated from z_0 by $\gamma_{b,c}$. In particular, $\gamma_{b,c} \in \Gamma_{a,c}$. In the third case, the same argument implies that $\gamma_{a,b} \in \Gamma_{a,c}$.

Summing up, we have that either one of $\gamma_{a,b} \cup \gamma_{b,c}$, $\gamma_{a,b}$, or $\gamma_{b,c}$ is in $\Gamma_{a,c}$. Since $\lambda(\Gamma_{a \cup b}) = \lambda(\Gamma_{b \cup c}) = 0$, by comparison, the extremal length of $\Gamma_{a,c}$ is also 0.

Definition 4.23. Let Ω be a simply connected domain. We define *prime ends* in Ω as equivalence classes of fundamental sequences in Ω.

Since this definition is given in terms of extremal length, we immediately get that prime ends are conformally invariant, namely, we get the following theorem which is a natural counterpart of Theorem 2.34.

Theorem 4.24. *Let (a_n) be a sequence in Ω and $f : \Omega \to \Omega'$ be a conformal map. Then (a_n) is fundamental if and only if $(f(a_n))$ is fundamental.*

This theorem gives a very indirect way to show that the new definition of prime ends is equivalent to the one given in Section 2.7.2. Both theorems give canonical correspondence between prime ends in Ω and in a reference domain, say $\mathbb{D}$. In both cases, prime ends in $\mathbb{D}$ are canonically equivalent to the boundary points. This gives the canonical bijection between two versions of prime ends. In the following, we give a more constructive proof.

The benefit of the new approach is that the proof of Theorem 4.24 is trivial, unlike the proof of Theorem 2.34 which requires non-trivial considerations and length–area arguments. The downside is that it is more difficult to work with cross-cut clusters than with cross-cuts. Fortunately for us, the difference between cross-cuts and cross-cut clusters is minor; moreover, it turns out that it is sufficient to consider the usual Euclidean metric only.

Theorem 4.25. *The sequence (a_n) is fundamental in Ω if and only if the reference point z_0 can be separated from almost all a_n by a cross-cut of an arbitrarily small diameter.*

Proof. Let us assume that $a = (a_n)$ is fundamental. Then there are clusters separating z_0 from almost all a_n. Take $\delta < d/4$, where $d = \operatorname{dist}(z_0, \partial\Omega)$ and consider a cluster γ with diameter at most δ. Let ζ be a point on γ and let us consider a circle of diameter 2δ centred at ζ. The intersection of this circle with Ω separates z_0 from γ. This intersection is made of at most countable union of circular arcs, and since γ is connected, one of these arcs separates γ from z_0. This implies that it also separates z_0 from almost all a_n. Since the diameter of this arc is at most 4δ, this implies that we can construct arbitrarily short cross-cuts separating z_0 from almost all a_n.

To prove the opposite, we assume that δ is much smaller than d and let γ be a cross-cut of diameter at most δ separating almost all a_n from z_0. Let ζ to be one of the endpoints of γ. Define c_r to be a circle of radius r centred at ζ. Let Γ_δ be the family of curves separating $c_{d/2}$ from $c_{2\delta}$ inside Ω. Any curve from this family contains a cross-cut separating γ from z_0, and hence it also separates z_0 from almost all a_n. By the comparison rule, $\lambda(\Gamma_a) \leq \lambda(\Gamma_\delta)$. Applying the comparison rule one more time, we get that $\lambda(\Gamma_\delta)$ is bounded by the extremal length of all curves in the annulus $\{z : 2\delta < |z - \zeta| < d/2\}$ that separate two boundary components. We already know that this extremal length is equal to $2\pi/\log(d/4\delta)$. This proves that $\lambda(\Gamma_a) \leq 2\pi/\log(d/4\delta)$ for any sufficiently small δ. This implies that $\lambda(\Gamma_a) = 0$ and the sequence (a_n) is fundamental. □

We say that a fundamental sequence (a_n) is equivalent to a null chain (γ_n) if for every reference point z_0 and sufficiently large n, every γ_n separates almost all a_m from z_0. With minimal modifications, the above theorem implies that this notion of equivalence respects equivalences for fundamental sequences and null chains. This means that the two notions of prime ends are equivalent.

Theorem 4.26. *For each null chain, there is an equivalent fundamental sequence and for each fundamental sequence there is an equivalent null chain. Moreover, for equivalent chains, the corresponding sequences are equivalent and for equivalent sequences the corresponding chains are equivalent. In other words, the equivalence of null chains and fundamental sequences extends to the equivalence between prime ends defined by null chains and prime ends defined by fundamental sequences.*

Proof. Let us denote the domain by Ω and fix a reference point z_0. Let (γ_n) be a null chain. Let us consider a sequence $a = (a_n)$ such that $a_n \in \gamma_n$ for every n. By construction, γ_n separates all a_m with $m > n$ from z_0. Since diameters of cross-cuts in a null chain tend to zero, by Theorem 4.25, the sequence (a_n) is fundamental. By definition, (a_n) is equivalent to (γ_n).

The other direction is more involved. From the proof of the previous theorem, we know that for a fundamental sequence $a = (a_n)$, we can construct arbitrarily short cross-cuts separating almost all a_n from a given reference point. Moreover, all cross-cuts are circular arcs. This is almost what we need, but it might happen that the corresponding sequence of cross-cuts is not a null chain; it could be that the cross-cuts are not separated. To construct a null chain, we have to modify the construction a little bit.

We will construct the null chain by induction. Let us assume that we have already constructed cross-cuts γ_m for $m \leq n$ such that their closures are disjoint and γ_k separates γ_{k+1} from z_0 for all $k < n$. Let c be any continuous curve connecting γ_n to z_0. By compactness, there is $\delta_n > 0$ such that the distance between c and $\partial\Omega$ is at least δ_n. This means that there is no cross-cut of diameter less than δ_n which separates γ_n from z_0. Let us choose $d_n > 0$ which is much smaller (say, by factor of 10) than δ_n, the distance between γ_n and γ_{n-1}, and 2^{-n}. Let us define γ_{n+1} to be a circular cross-cut of diameter at most d_n which separates almost all a_k from z_0. The existence of such a cut is given by Theorem 4.25. Since the diameter of this cut is much smaller than δ_n, it cannot be outside of γ_n. This means that it is either strictly inside or their closures intersect. In the first case, there is nothing left to do. In the second case, we have to carry out some modifications. Let us consider c_n — a circular cross-cut with the same centre as γ_{n+1} but with radius twice as long. This cross-cut also separates almost all a_k from z_0. We modify γ_n to be a cross-cut which is made of parts of γ_n and c_n (see Figure 4.10). Since the diameter of c_n is much smaller than the

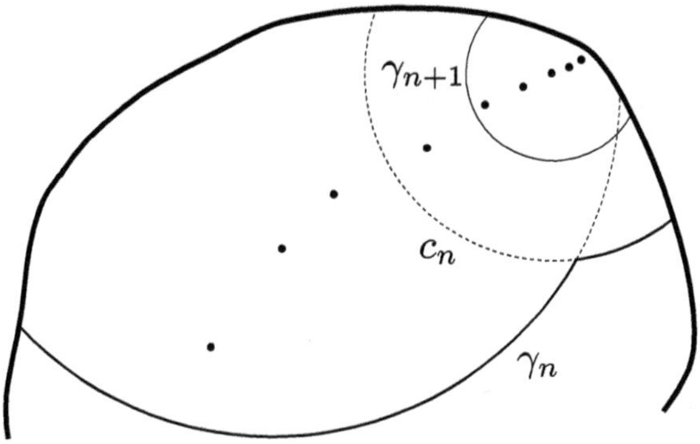

Fig. 4.10. Curve c_n cannot be outside of γ_n; hence they must intersect. We modify γ by taking the union of a part of c_n and a part of γ_n.

distance between γ_{n-1} and the original γ_n, the new γ_n is still well separated from γ_{n-1}. Continuing in this way, we construct a null chain equivalent to (a_n).

Finally, let us show that the equivalence of null chains and fundamental sequences respect the corresponding equivalences. Let (γ_n) be a null chain, let $a = (a_n)$ and $a' = (a'_n)$ be two fundamental sequences that are equivalent to (γ_n). For any $\delta > 0$, there is γ_n with diameter smaller than δ. This curve separates almost all a_n and a'_n from z_0, hence it separates almost all elements of $a \cup a'$ from z_0. By Theorem 4.25, this implies that $a \cup a'$ is fundamental and that a is equivalent to a'.

Let $a = (a_n)$ be a fundamental sequence, and (γ_n), (γ'_n) be two null chains equivalent to a. We claim that these null chains are equivalent. Let γ_n be any cross-cut from the first chain. By the same argument as above, there is $\delta_n > 0$ such that no cross-cut separating almost all a_n from z_0 could lie outside of γ_n. Since the diameters of γ'_n tend to zero, either there is $N = N(n)$, such that for all $m > N$ the curve γ'_m is inside of γ_n, or almost all γ'_m intersect γ_n. We claim that the second alternative cannot happen. Let us consider γ_{n+1} which is a fixed distance away from γ_n; if the diameter of γ'_m is smaller than this distance, then γ'_m cannot intersect both γ_n and γ_{n+1}. Since they intersect γ_n, this implies that they must be outside of γ_{n+1}, which is impossible since both of them separate almost all a_n from z_0. This proves that γ_n separates almost all γ'_m from z_0. Applying the same argument to γ'_n, we show that null chains are equivalent. □

4.5. Solutions

Solution of 4.1. Let T be the symmetry with respect to the unit circle, namely $T(z) = 1/\bar{z}$. Let Ω be symmetric with respect to T, that is, $z \in \Omega$ if and only if $T(z) \in \Omega$. Let Γ be a symmetric family of curves, namely $\gamma \in \Gamma$ if and only if $T(\gamma) \in \Gamma$. Then

$$\lambda(\Gamma) = \sup_{\rho} \frac{L^2(\Gamma, \rho)}{A(\Omega, \rho)},$$

where supremum is over all symmetric admissible metrics ρ, that is, metrics ρ such that $\rho(z) = \rho(T(z))/|z|^2$ (the factor $|z|^{-2}$ is the Jacobian of T).

The proof is essentially the same as in the case of the symmetry with respect to the real line. For a metric ρ_1, we define $\rho_2(z) = \rho_1(T(z))/|z|^2$. By the change of variable formula and symmetry of Γ, we have $L(\Gamma, \rho_1) = L(\Gamma, \rho_2)$. By the same argument $A(\Omega, \rho_1) = A(\Omega, \rho_2)$. Let us consider the metric $\rho = (\rho_1 + \rho_2)/2$. By construction, this metric is symmetric and as before

$$\frac{L^2(\Gamma, \rho)}{A(\Omega, \rho)} \geq \frac{L^2(\Gamma, \rho_1)}{A(\Omega, \rho_1)} = \frac{L^2(\Gamma, \rho_2)}{A(\Omega, \rho_2)},$$

which immediately implies that it is sufficient to take supremum over symmetric metrics. □

Solution of 4.2. By symmetry with respect to the y-axis, it is obvious that the extremal distance between $[-1, 0]$ and $[1, +\infty]$ is the same as the extremal distance between $[-\infty, -1]$ and $[0, 1]$. But they are conjugated and their product is 1. This proves that they are both equal to 1. Let R be a rectangle conformally equivalent to our configuration. By conformal invariance, the extremal distance between its opposite sides is 1. Since the extremal distance in the rectangle is equal to the ratio of sides, the rectangle must be a square. □

Solution of 4.3. It is easy to see that $d_{\Omega_1}(E_1, F_1) = d_{\Omega_2}(E_2, F_2)$. Indeed, there is one-to-one correspondence between the corresponding families of curves and metrics. For a metric $\rho_1(z)$ in Ω_1, we define $\rho_2(z) = \rho_1(\bar{z})$ in Ω_2. Moreover, all metrics in Ω_2 can be written this way. For each γ_1 connecting E_1 and F_1 in Ω_1, the symmetric curve $\bar{\gamma}_1$ connects E_2 and F_2 in Ω_2. Since $L(\gamma_1, \rho_1) = L(\bar{\gamma}_1, \rho_2)$, we have that the extremal distances are equal.

Next, we consider the family of curves that correspond to $d_\Omega(E, F)$. By the symmetry rule, it is sufficient to consider only symmetric metrics

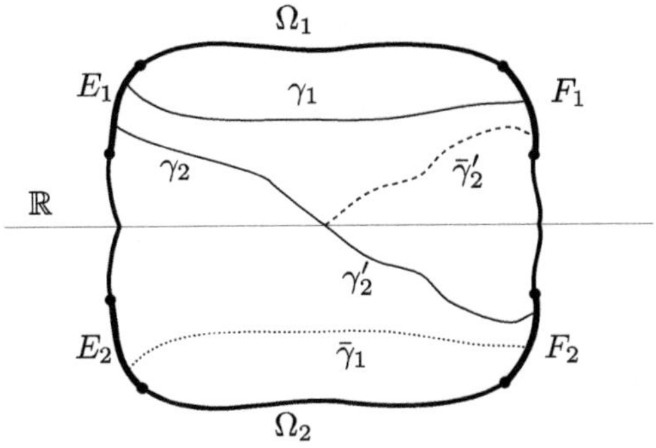

Fig. 4.11. Application of symmetry rule curve connecting E_1 with F_2 could be transformed into a curve of the same length which connects E_1 and F_1.

ρ in Ω. For curve γ_2 connecting E and F, we can construct a curve in Ω_1 connecting E_1 and F_1, namely, we replace curve $\gamma_2(t) = (x(t), y(t))$ by $(x(t), |y(t)|)$; see Figure 4.11. This implies that $L(\Gamma, \rho) = L(\Gamma_1, \rho)$, where Γ is the family of curves defining $\mathrm{d}_\Omega(E, F)$ and Γ_1 is the family of curves defining $\mathrm{d}_{\Omega_1}(E_1, F_1)$. Since, obviously, $A(\Omega, \rho) = 2A(\Omega_1, \rho)$ we have that $2d_\Omega(E, F) = d_{\Omega_1}(E_1, F_1)$. □

Solution of 4.4. To obtain the lower bound, we could consider $\rho = 1$, that is, the usual Euclidean metric. It is clear that $L(\Gamma, \rho) = 2$ and $A(\Omega, \rho) = 4(1 - \varepsilon^2)$. This gives the lower bound $4/(4(1 - \varepsilon^2)) = 1/(1 - \varepsilon^2)$.

To obtain the upper bound, we consider Γ', the family of curves connecting top to bottom within $([-1, -\varepsilon] \cup [\varepsilon, 1]) \times [-1, 1]$. Each curve in this family is also a curve in Γ. By the comparison rule, $\lambda(\Gamma) \leq \lambda(\Gamma')$. By symmetry, $\lambda(\Gamma') = \lambda(\Gamma'')/2$, where Γ'' is the family of curves connecting top to bottom in $[\varepsilon, 1] \times [-1, 1]$. The last extremal distance is the modulus of a rectangle and is equal to $2/(1 - \varepsilon)$. Combining these results, we get $\lambda(\Gamma) \leq 1/(1 - \varepsilon)$. □

Solution of 4.5. It is sufficient to compute the extremal distance between $E' = \{e^{i\theta}, 0 \leq \theta \leq \theta_0\}$ and $[-1, 0]$ in the upper half-disc. The function $J(z) = z/2 + 1/2z$ conformally maps the upper half-disc onto the lower half-plane in such a way that $J([-1, 0]) = [-\infty, -1]$ and $J(E') = [\cos(\theta), 1]$. The simplest geometry where we can compute the extremal distance is the rectangle, where it is the ratio of side lengths. To map it to the rectangle, we first find a Möbius transformation that sends ∞, -1, $\cos(\theta)$, and 1 to $-1/k$, -1, 1, and $1/k$. Möbius transformations preserve cross-ratios. This means that such a map exists if and only if cross-ratios are the same. This gives us a relation between $\cos(\theta)$ and k:

$$\frac{2}{1+\cos(\theta)} = \frac{(1+k)^2}{4k}$$

or (since $0 < k < 1$),

$$k = k(\theta) = \frac{\sqrt{2} - \sqrt{1-\cos(\theta)}}{\sqrt{2} + \sqrt{1-\cos(\theta)}}.$$

The Christoffel–Schwarz function maps $\mathbb{H}$ with these marked points onto a rectangle with the ratio of side lengths

$$\frac{2K(k)}{K'(k)}.$$

Hence, $\lambda(G) = 2K(k)/K'(k)$, where K and K' are elliptic integrals and $k = k(\theta)$. □

Solution of 4.6. Let Γ be the family of all loops separating z_1 and z_2 from the boundary of Ω. We have shown that the extremal length of this family can be written as an explicit function of $G_\Omega(z_1, z_2)$. It is easy to check that this function is monotone. But this family does not change if we relabel the points, which proves that Green's function is symmetric. □

Solution of 4.7. By conformal invariance, we assume that $\Omega = \mathbb{D}$, $z_1 = 0$ and $z_2 = x \in (0,1)$. By symmetry of this configuration, it is sufficient to consider symmetric metrics ρ. By a standard symmetrization argument, for a symmetric metric ρ and every curve $\gamma \in \Gamma^*$, there is a symmetric curve $\tilde{\gamma}$ which stays in the upper half-plane, connects the unit circle and the interval $[0, x]$ such that $L(\tilde{\gamma}, \rho) \leq L(\gamma, \rho)/2$. This implies that $\lambda(\Gamma^*)$ is twice the extremal length of the family of such curves $\tilde{\gamma}$ (the other inequality follows from the comparison rule). The extremal length of this family is the extremal distance between the unit circle and $[0, x]$ inside the upper half-disc. We have seen before that $\lambda(\Gamma)$ is twice the extremal distance between $[-1, 0]$ and $[x, 1]$ in the upper half-disc. Since the product of these two extremal distances is 1, we have that $\lambda(\Gamma)\lambda(\Gamma^*) = 4$. □

Solution of 4.8. By applying an appropriate linear transformation, we can assume, without loss of generality, that the inner component is of diameter 1 and it contains both -1 and 0. We claim that the unit circle is inside Ω. Let us assume that this is not the case; then there is a point in the outer component with modulus less than 1. By the Teichmüller theorem, the modulus of Ω is smaller than the modulus of the Teichmüller annulus Ω_R with $R = 1$. What is left is to show that the modulus of Ω_1 is $1/2$.

So, we have to compute the extremal distance between $[-1, 0]$ and $[1, +\infty]$ in $\mathbb{C}$. By a version of the symmetry rule given in Exercise 4.3, this extremal distance is equal to $d_{\mathbb{H}}([-1, 0], [1, +\infty])/2$. By Exercise 4.2, $d_{\mathbb{H}}([-1, 0], [1, +\infty]) = 1$, and hence the modulus of the Teichmüller annulus Ω_1 is equal to $1/2$. □

Solution of 4.9. By conformal invariance, we can assume that Ω is an equilateral triangle with vertices 0, 1, and $1/2 + i\sqrt{3}/2$. Let us consider $\Gamma_0 = \Gamma_1 \cup \Gamma_2 \cup \Gamma_3$, where Γ_1 is the family of curves γ_x that start at $x \in [1/2, 1]$ (the right half of the horizontal side), go vertically to the right side, and then go perpendicularly to the left side of the triangle. See Figure 4.12 for an example of such a curve. Families Γ_2 and Γ_3 are defined in the same way, but start on the right and left sides, respectively. Let ρ be equal to 1 inside the triangle and 0 outside. We claim that by Theorem 4.4, this metric is the extremal. We have to check that ρ and Γ_0 satisfy the assumption of this theorem.

The length of any curve $\gamma_1 \in \Gamma_1$ is the same as the length of a vertical interval connecting the bottom side with the horizontal side passing through the top vertex. This proves that all curves in Γ_0 have the same length $\sqrt{3}/2$. Essentially, the same argument shows that any curve going from the bottom side to the right and then to the left side has length at least $\sqrt{3}/2$. This proves that Γ_0 is the family of shortest curves in the metric ρ.

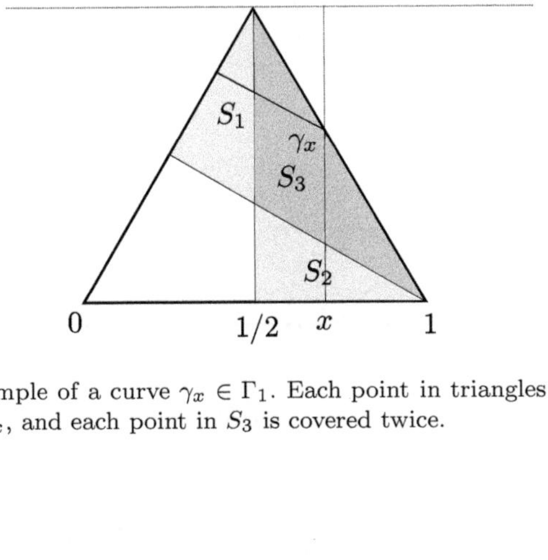

Fig. 4.12. An example of a curve $\gamma_x \in \Gamma_1$. Each point in triangles S_1 and S_2 belongs to exactly one of γ_x, and each point in S_3 is covered twice.

Let us assume that $\int_\gamma h \geq 0$ for some function h and every $\gamma \in \Gamma_0$. Let us assume that $\gamma = \gamma_x \in \Gamma_1$. Integrating with respect to x, we get

$$0 \leq \int_{1/2}^{1} \int_{\gamma_x} h\rho \, \mathrm{d}s \mathrm{d}x = \int_{S_1} h\rho + \int_{S_2} h\rho + 2\int_{S_3} h\rho,$$

where S_1 and S_2 are light grey triangles and S_3 is the darker one (see Figure 4.12). We have the factor 2 in front of the last integral since each point in S_3 is covered by two curves. Summing similar formulas for Γ_2 and Γ_3, we get

$$3\int_\Omega \rho h \geq 0.$$

This proves that we can apply Theorem 4.4 to show that $\rho = 1$ is indeed extremal. For this measure, $L(\Gamma, \rho) = \sqrt{3}/2$ and $A(\Omega, \rho) = \sqrt{3}/4$. This proves that $\lambda(\Gamma) = \sqrt{3}$. □

Solution of 4.10. It is important to note that this problem is different from the one in Exercise 4.9. The family Γ is not symmetric with respect to relabelling vertices and it is easy to show that if Ω is the equilateral triangle and $\rho = 1$, then this ρ is not an extremal metric.

In this case, we can use the triangle with vertices $a = (-1, 0)$, $b = (1, 0)$, and $c = (0, 1)$. Take $\rho = 1$ in this domain and define Γ_0 to be the family of all curves γ_x that are made of two intervals orthogonal to the left and right sides and meeting at x on the horizontal side (see Figure 4.13). We claim that this metric is extremal. The proof follows the same lines as all other proofs that a metric is extremal. First, we note that all these curves have length $\sqrt{2}$ and it is easy to see that they are the shortest curves with

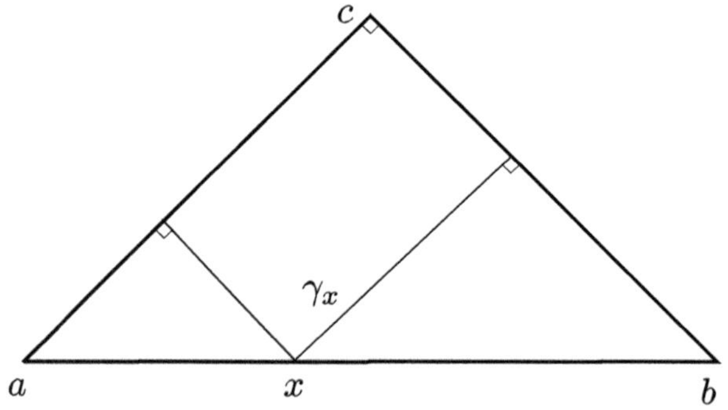

Fig. 4.13. An example of a curve $\gamma_x \in \Gamma$.

respect to $\rho = 1$. If $\int_\gamma h \geq 0$ for each γ_x then, integrating with respect to x, we get that the integral of h over the triangle is also non-negative. By Theorem 4.4, this proves that $\rho = 1$ is extremal. Since the area of this triangle is 1, we have that $\lambda(\Gamma) = (\sqrt{2})^2/1 = 2$. □

Solution of 4.11. By conformal invariance, we can assume that $z_1 = -1$ and $z_2 = 1$. In this case, Γ is symmetric with respect to the real line and it is sufficient to consider only metric ρ that is symmetric with respect to the real line.

Using the same symmetrization as in Figure 4.8, we show that it is sufficient to consider the family of symmetric curves γ such that their top part $\gamma_+ \subset \bar{\mathbb{H}}$ connects $(-\infty, -1)$ to $(-1, 1)$ and then to $(1, \infty)$. Let us denote the family of such curves by Γ_+.

By symmetry, $\lambda(\Gamma) = 2\lambda(\Gamma_+)$. It is easy to see that the family Γ_+ is conformally invariant to the family considered in Exercise 4.10. This proves that $\lambda(\Gamma) = 4$. □

Solution of 4.12. We know that the conformal map $f : \Omega \to \mathbb{H}$ is given by $\sqrt{z^2+1}$. For this map, $f(\zeta) = 0$, $f(z_0) = i\sqrt{3}$. The r-neighbourhood of ζ is $[i - ir, i]$ and its image is $[-\sqrt{2r - r^2}, \sqrt{2r - r^2}]$. As $r \to 0$, this interval shrinks to 0 and its harmonic measure divided by its length l_r converges to the Poisson kernel which is equal to $1/\pi\sqrt{3}$. The length l_r is equal to $2\sqrt{2r - r^2}$ and clearly $l_r/2\sqrt{2r} \to 1$. Combining all these asymptotics together, we have

$$\omega(z_0, B(\zeta, r)) \approx \frac{1}{\pi\sqrt{3}} 2\sqrt{2r},$$

where $\approx$ means that the ratio tends to 1 as $r \to 0$. □

Solution of 4.13. Let Ω be a polygonal domain such that 0 is one of its vertices and the angle at this vertex is $\pi\alpha$. From Theorem 2.17, we know that a conformal map from $\mathbb{H}$ onto Ω is given by

$$C \int_0^z z^{-\beta} \prod (z - x_i)^{-\beta_i} \mathrm{d}z = \int_0^z z^{-\beta} g(z) \mathrm{d}z,$$

where $\beta = 1 - \alpha$ and $g(z)$ is the product of all factors except $z^{-\beta}$.

The function g is continuous near $z = 0$ and $g(0) \neq 0$. This implies that near the origin $f(z) \approx z^\alpha g(0)/\alpha = cz^\alpha$. This is something that we should expect: the map $z \mapsto z^\alpha$ maps the half-plane onto a wedge with angle $\pi\alpha$. Our computation means that the local behaviour near the corner is the same.

The preimage of an r neighbourhood of the origin is an interval which has the length of order $r^{1/\alpha}$ and its harmonic measure is comparable to its length (independently of the position of z_0 — the pole of harmonic measure). Hence,

$$\omega(B(\zeta, r)) \approx cr^{1/\alpha},$$

where c depends on z_0 and the shape of Ω, but not on r. This proves that the dimension of harmonic measure at the corner is independent of z_0 and is equal to $1/\alpha$. Since α can be anything between 0 and 2, we have that the dimension can take any value between 1/2 and infinity. □

Chapter 5

Loewner Evolution

Loewner evolution is a classical subject first introduced in 1923 by Charles Loewner[1] in his famous paper, "Untersuchungen über schlichte konforme Abbildungen des Einheitskreises. I". This work was motivated by the extremal problems of function theory and showed a way to describe an evolution of slit maps. The first application of this technique was the proof of the $n = 3$ case of the Bieberbach conjecture. Bieberbach, who was the editor of *Mathematische Annalen*, the journal where Loewner published his paper, was so sure that this technique was sufficiently powerful that Loewner would be able to give the complete proof very soon that he added 'Part I' to the title. But Loewner never returned to the subject and to some extent the Loewner evolution had faded away until interest was revived in 1985 when de Branges (1985) used it to finally prove the Bieberbach conjecture. After the initial excitement which revived interest in the Loewner evolution, it went into decline one more time, but interest was renewed when it became one of the key ingredients in the seminal paper of Oded Schramm (2000). In this paper, he introduced a stochastic version of the Loewner evolution which is called SLE. Originally, SLE meant stochastic Loewner evolution, but in recent years, people have started to use Schramm–Loewner evolution instead. This stochastic version turned out to be an extremely valuable tool in the study of critical models of statistical

[1] Loewner used several spellings of his name: he was born as Karel Löwner, but changed his first name to Karl after moving from Prague to Berlin. In 1939, he immigrated to the USA where he anglicized his name. Nowadays, he is mostly known as Charles Loewner and we will use this spelling throughout the book.

physics and led to a very rapid progress in the mathematically rigorous understanding of these models. Interested readers can find more information about SLE in books by Lawler (2005) and Kemppainen (2017). This book also contains an introduction to the theory of Loewner evolution written from a more probabilistic perspective.

5.1. Basic Loewner Evolutions for Slit Maps

5.1.1. *Radial Loewner evolution*

We start by considering the basic setup: Let Ω be a simply connected domain containing the origin and $\gamma(t)$ be a Jordan curve growing from a point on the boundary of Ω towards a point inside. We also assume that γ avoids the origin.

To be more precise, we define $\gamma(t)$ to be a continuous injective function on $[0, T]$ such that $\gamma(0) \in \partial\Omega$ and $\gamma((0, T]) \subset \Omega$. Moreover, we assume that $\gamma(t) \neq 0$ for all $t \in [0, T]$. We do not assume that Ω is bounded, but we assume that the starting point $\gamma(0)$ is finite. This assumption is not essential and will be dropped out later, but it is convenient for initial consideration. In this case, we can define $\Omega_t = \Omega \setminus \gamma([0, t])$, which is clearly a simply connected domain containing the origin. By the Riemann mapping theorem, we can uniquely define a conformal map

$$f_t(z) : \mathbb{D} \to \Omega_t,$$

such that $f_t(0) = 0$ and $f_t'(0) > 0$. By $g_t(z)$, we denote the inverse map $f_t^{-1}(z)$. Our main goal is to study how the maps f_t and g_t change with time. Throughout this chapter, we will always use f to denote maps from a uniformizing domain and g to denote maps onto a uniformizing domain.

Remark 5.1. This setup is slightly different from the one which can be found in many classical books. In our case, we are *growing* a curve starting from a point on the boundary. The resulting domains are *decreasing*. Many authors prefer to start with a slit and *erase* it starting from its tip. This results in an *increasing* family of domains. The difference between two approaches is purely cosmetic and corresponds to the time reversal. The resulting equations are slightly different: they differ by changing $t \mapsto -t$. We use this convention since it is the standard convention in the modern study of the Loewner evolution, especially when it comes to the Schramm–Loewner evolution.

From the continuity of γ, one can immediately derive that the family of domains Ω_t is continuous in the sense of the kernel or Carathéodory convergence (with respect to the origin). In particular, this means that the family (f_t) is continuous with respect to the uniform convergence on compact sets. Functions f_t can be expanded near the origin as

$$f_t(z) = a_1(t)z + a_2(t)z^2 + \cdots,$$

where $a_n(t)$ are continuous functions of t. Recall that the first constant $a_1(t)$ is positive and equal to the conformal radius of Ω_t. Since $\Omega_t \subsetneq \Omega_s$ for $0 \le s < t \le T$, by Lindelöf's principle, $a_1(t) < a_1(s)$. Alternatively, we can define the *transition function*

$$\phi_{s,t}(z) = f_s^{-1}(f_t(z)) = g_s(f_t(z)),$$

which is an analytic function in $\mathbb{D}$ such that $\phi_{s,t}(0) = 0$ and $|\phi_{s,t}(z)| \le 1$. By the Schwarz lemma, $\phi'_{s,t}(0) = a_1(t)/a_1(s) \le 1$. Equality happens if and only if $\phi_{s,t}(z) = z$, which is impossible since in this case, $\Omega_s = \Omega_t$. This proves that the function $a_1(t)$ is strictly decreasing.

Particular parametrization of the curve γ is not very important for our purposes; we are mostly interested in the properties that do not change under reparametrization. Since the function a_1 is strictly decreasing, we can reparametrize γ in such a way that $a_1(t) = \beta e^{-t}$, where $\beta = a_1(0)$. From now on, unless explicitly stated otherwise, this will be the standard parametrization.

Note that since $\gamma([0, t])$ is a nice Jordan curve, we can define $\lambda(t)$ as the unique point on the unit circle such that $f_t(\lambda(t)) = \gamma(t)$. Our first claim is that $\lambda(t)$ is a continuous function of t (see Figure 5.1).

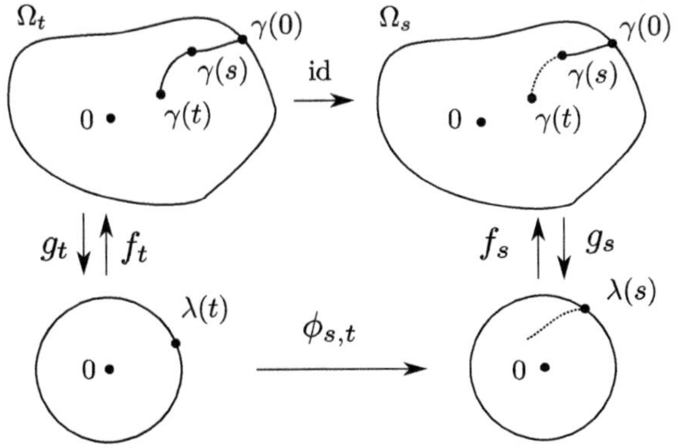

Fig. 5.1. Increment of a slit map. Function $\phi_{s,t}$ maps the unit disc onto the unit disc with a slit from $\lambda(s)$ which is the image of $\gamma(s)$ under g_s.

Lemma 5.2. *The function* $\lambda(t)\ :\ [0,T] \to \partial(\mathbb{D})$ *defined above is continuous.*

Proof. We start by introducing some additional notation. We denote $g_s(\gamma([s,t]))$ by $\gamma_{s,t}$. Note that $\phi_{s,t}$ is the Riemann map from $\mathbb{D}$ onto $\mathbb{D}\setminus\gamma_{s,t}$. We denote the arc of the unit circle which is the preimage of $\gamma_{s,t}$ under $\phi_{s,t}$ by $A_{s,t}$. The arc of the unit circle complementary to $A_{s,t}$ is mapped onto $\partial\mathbb{D}\setminus\lambda(s)$ by $\phi_{s,t}$. By the Schwarz reflection, $\phi_{s,t}$ can be extended to a map from $\mathbb{C}\setminus A_{s,t}$ onto $\mathbb{C}\setminus(\gamma_{s,t}\cup\tilde{\gamma}_{s,t})$, where $\tilde{\gamma}_{s,t}$ is the symmetric image of $\gamma_{s,t}$ under reflection with respect to the unit circle. We denote this extension by $\phi_{s,t}$ as well (see Figure 5.2).

We claim that as s and t are getting closer, the maps $\phi_{s,t}$ converge to the identity and that this proves the continuity of λ.

To prove convergence, we consider a family of analytic maps $\psi_{s,t} = \phi_{s,t}(z)/z$. These maps are also analytic in $\mathbb{C}\setminus A_{s,t}$ since the singularity at $z=0$ is removable. They are uniformly bounded at infinity for $s < t \le T$ since

$$\lim_{z\to\infty}\psi_{s,t}(z) = \lim_{z\to\infty}\frac{\phi_{s,t}(z)}{z} = \lim_{z\to 0}\frac{z}{\bar{\phi}_{s,t}(\bar{z})} = \frac{1}{\phi'_{s,t}(0)} = e^{t-s}, \tag{5.1}$$

where the last equality follows from the definition of $\phi_{s,t}$ and time parametrization. By the Koebe 1/4 theorem, the curve $\gamma_{s,t}$ is at least $e^{s-t}/4$ away from the origin. This implies that $\tilde{\gamma}_{s,t}$ is inside the disc of radius $4e^{t-s}$ and that $|\psi_{s,t}(z)| \le 4e^{t-s}$ on $A_{s,t}$. These two estimates, together with the maximum modulus, imply that $\{\psi_{s,t}, 0\le s<t\le T\}$ is a normal family.

Let us fix t and consider any sequence $s_n \nearrow t$ together with the corresponding sequence $\psi_{s_n,t}$. Since the sequence is normal, we can choose a subsequence converging to a bounded function ψ which is analytic in $\mathbb{C}\setminus\{\lambda(t)\}$. The point $\lambda(t)$ is an isolated singularity and ψ is bounded; hence it is a removable singularity and ψ is a bounded entire function, which implies that it is a constant. By (5.1), this constant must be 1. Since this is true for every sequence $s_n \nearrow t$, the same is true for $s \nearrow t$. This implies that $\phi_{s,t}(z) \to z$ uniformly on every compact in $\mathbb{C}\setminus\{\lambda(t)\}$. It is easy to believe that this statement implies that $\lambda(s)$ converges to $\lambda(t)$, but the rigorous proof is surprisingly not so simple.

As $s \nearrow t$, the curve $\gamma([s,t])$ shrinks to a point and so does $\gamma_{s,t}$. This implies that $A_{s,t}$ also shrinks to a point. Fix small $\varepsilon > 0$. Then the argument above implies that $A_{s,t}$ lies inside a circle of radius ε centred at $\lambda(t)$ for all s sufficiently close to t. We denote this circle by S_1 and let $S_2 = \phi_{s,t}(S_1)$ be its image which encloses $\gamma_{s,t}$ and $\tilde{\gamma}_{s,t}$ (see Figure 5.2).

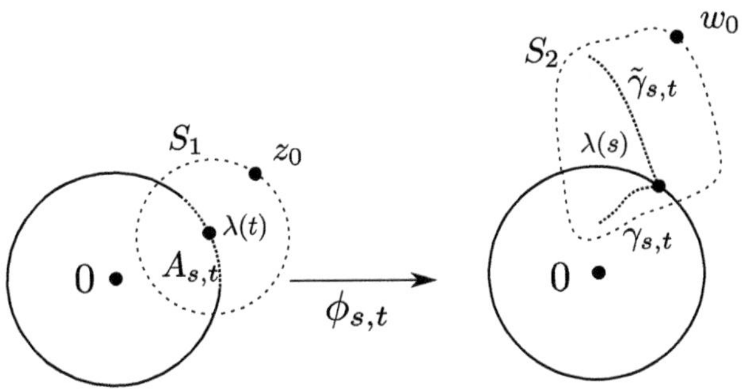

Fig. 5.2. The extended function $\phi_{s,t}$ is continuous up to the boundary and maps the (dashed) arc $A_{s,t}$ onto the union of (dashed) curves $\gamma_{s,t}$ and $\tilde{\gamma}_{s,t}$. The circle S_1 is mapped onto a closed curve S_2 which encloses $\gamma_{s,t}$ and $\tilde{\gamma}_{s,t}$.

Functions $\phi_{s,t}(z)$ converge to z uniformly on S_1, and hence, for s sufficiently close to t, $|\phi_{s,t}(z) - z| < \varepsilon$ on S_1. Let z_0 be some point on S_1 and w_0 be its image. Then we can write

$$|\lambda(t) - \lambda(s)| \leq |\lambda(t) - z_0| + |z_0 - w_0| + |w_0 - \lambda(s)|.$$

The first term is equal to ε and the second is bounded by ε. To estimate the last one, we note that S_1 has diameter 2ε and under $\phi_{s,t}$, all points move by at most ε; this means that all points on S_2 are at most 4ε apart from each other. Hence every point inside S_2 is also at most 4ε apart from any given point on S_2. This proves that for s sufficiently close to t, we have $|\lambda(s) - \lambda(t)| \leq 6\varepsilon$. This proves that λ is left-continuous.

Proof of right continuity is essentially the same, but we consider $\phi_{s,t}^{-1}$ instead of $\phi_{s,t}$. Alternatively, one can use the Beurling inequality (Theorem 4.21); see also the argument in Section 5.1.2. □

Now we are ready to prove our first version of the Loewner differential equation, also known as the Loewner evolution. This version is called the radial Loewner evolution since it describes the evolution of a curve growing from the boundary towards the origin.

Remark 5.3. In the following theorem, as well as in the rest of this chapter, we will use derivatives of functions like $f_t(z)$ and $g_t(z)$. We always consider z as a spatial variable and t as the time. We will denote the derivatives with respect to z by $f_t'(z)$ and derivatives with respect to t by $\dot{f}_t(z)$ or by $\partial_t f_t$.

Theorem 5.4. *Let f_t, g_t, and $\lambda(t)$ be as above. Then $g_t(z)$ satisfies an ordinary differential equation*

$$\dot{g}_t(z) = g_t(z)\frac{\lambda(t) + g_t(z)}{\lambda(t) - g_t(z)}, \tag{5.2}$$

with the initial condition $g_0(z) = g(z)$, where $g(z) : \Omega \to \mathbb{D}$ is the Riemann map with $g(0) = 0$ and $g'(0) > 1$. The family f_t satisfies a partial differential equation

$$\dot{f}_t(z) = -zf_t'(z)\frac{\lambda(t) + z}{\lambda(t) - z}, \tag{5.3}$$

with the initial condition $f_0(z) = f(z)$, where $f = g^{-1}$.

One can assume that families of maps f_t and g_t are solutions of partial or ordinary differential equations that consider $\lambda(t)$ as an 'input signal'. For this reason, we call $\lambda(t)$ the *driver* or the *driving function* of the Loewner evolution. This way of approaching the Loewner evolution will be studied in more detail later.

Proof. Let us consider the functions $\psi_{s,t}(z) = \phi_{s,t}(z)/z$ that we have introduced before. Function $\phi_{s,t}$ is equal to 0 only at $z = 0$ where it has a simple zero; this proves that $\psi_{s,t} \neq 0$ and it is possible to define in $\mathbb{D}$ a single-valued function

$$\Psi_{s,t}(z) = \log\left(\frac{\phi_{s,t}(z)}{z}\right) = \log\psi_{s,t}(z).$$

We choose the branch of logarithm such that $\Psi_{s,t}(0) = s - t$. Note that $\Psi_{s,t}$ is analytic in $\mathbb{D}$ and continuous up to the boundary.

Using the Schwarz formula (2.7), we can write $\Psi_{s,t}$ in terms of its real part on the boundary:

$$\Psi_{s,t}(z) = \frac{1}{2\pi}\int_0^{2\pi} \mathrm{Re}\left(\Psi_{s,t}(e^{i\theta})\right)\frac{e^{i\theta} + z}{e^{i\theta} - z}\mathrm{d}\theta + iC.$$

By considering $z = 0$, we have

$$s - t = \Psi_{s,t}(0) = \frac{1}{2\pi}\int_0^{2\pi} \mathrm{Re}\left(\Psi_{s,t}(e^{i\theta})\right)\mathrm{d}\theta + iC.$$

From this formula, we immediately have that $C = 0$ and the average value of $\mathrm{Re}\,\Psi_{s,t}$ on the boundary is equal to $s-t$. We also note that on $\partial\mathbb{D}\setminus A_{s,t}$, the modulus of $\psi_{s,t}$ is equal to 1, so $\operatorname{Re}\Psi_{s,t} = 0$ on the arc complementary to $A_{s,t}$. Thus, slightly abusing notation, the formulas above can be rewritten as

$$\Psi_{s,t}(z) = \frac{1}{2\pi}\int_{A_{s,t}} \operatorname{Re}(\Psi_{s,t}(e^{i\theta}))\frac{e^{i\theta}+z}{e^{i\theta}-z}\mathrm{d}\theta \tag{5.4}$$

and

$$s - t = \frac{1}{2\pi}\int_{A_{s,t}} \operatorname{Re}(\Psi_{s,t}(e^{i\theta}))\mathrm{d}\theta. \tag{5.5}$$

By the definition of $\phi_{s,t}$ (see also Figure 5.1), we have

$$g_s(z) = \phi_{s,t}(g_t(z)).$$

Using this identity and plugging $g_t(z)$ instead of z into (5.4), we obtain

$$\log\left(\frac{g_s(z)}{g_t(z)}\right) = \Psi_{s,t}(g_t(z)) = \frac{1}{2\pi}\int_{A_{s,t}} \operatorname{Re}(\Psi_{s,t}(e^{i\theta}))\frac{e^{i\theta}+g_t(z)}{e^{i\theta}-g_t(z)}\mathrm{d}\theta.$$

This integral can be split into real and imaginary parts. Since $\operatorname{Re}(\Psi_{s,t}) < 0$ on $A_{s,t}$, we can apply the mean value theorem to both real and imaginary parts. The result is that there are θ_R and θ_I that belong to $A_{s,t}$ such that

$$\begin{aligned}
\log(g_s(z)) - \log(g_t(z)) &= \log\left(\frac{g_s(z)}{g_t(z)}\right)\\
&= \frac{1}{2\pi}\left[\operatorname{Re}\frac{e^{i\theta_R}+g_t(z)}{e^{i\theta_R}-g_t(z)} + i\operatorname{Im}\frac{e^{i\theta_I}+g_t(z)}{e^{i\theta_I}-g_t(z)}\right]\\
&\quad\times\int_{A_{s,t}} \operatorname{Re}(\Psi_{s,t}(e^{i\theta}))\mathrm{d}\theta\\
&= \left[\operatorname{Re}\frac{e^{i\theta_R}+g_t(z)}{e^{i\theta_R}-g_t(z)} + i\operatorname{Im}\frac{e^{i\theta_I}+g_t(z)}{e^{i\theta_I}-g_t(z)}\right](s-t),
\end{aligned}$$

where the last equality follows from (5.5). Dividing by $(s-t)$ and passing to the limit as $s \nearrow t$ and using that $A_{s,t}$ shrinks to $\lambda(t)$, we prove that

$$\frac{\dot{g}_t(z)}{g_t(z)} = \partial_t \log(g_t(z)) = \frac{\lambda(t)+g_t(z)}{\lambda(t)-g_t(z)},$$

which is equivalent to equation (5.2). Since we considered the limit as $s \nearrow t$, we have shown that the left derivative of g_t satisfies (5.2). A similar argument shows that the right derivative also exists and satisfies the same equation. This proves that $\dot{g}_t(z)$ exists and satisfies (5.2).

Assuming that f_t is differentiable with respect to t, we can derive (5.3) from (5.2). Indeed, by definition, $f_t(g_t(z)) = z$. Differentiating this with respect to t, we arrive at $\dot{f}_t(g_t(z)) + f_t'(g_t(z))\dot{g}_t(z) = 0$. Applying (5.2) and replacing $g_t(z)$ by z, we obtain (5.3).

It remains to be shown that in the above argument, we can indeed differentiate with respect to t. To show this, let us consider $F(z,t) = (f_t(z), t)$ which is a function from $\mathbb{D} \times [0,T]$ to $\mathbb{C} \times [0,T]$. This function is the inverse of $G(z,t) = (g_t(z), t)$ which is continuously differentiable. Its Jacobian matrix is

$$\begin{bmatrix} g_t'(z) & \dot{g}_t(z) \\ 0 & 1 \end{bmatrix}.$$

Its determinant is $g_t'(0)$, which never vanishes. This means that we can apply the inverse function theorem and conclude that F is continuously differential as well, which immediately implies that f_t is continuously differential with respect to t. □

Remark 5.5. As mentioned above, functions f_t have expansion $f_t(z) = a_1(t)z + \cdots$ at the origin and we work with the normalization $a_1(t) = \beta e^{-t}$. The equation for the general case could be extracted from the explicit time change formula. Alternatively, we can note that $(s-t)$ in formula (5.5) is equal to $\log(a_1(t)/a_1(s))$, so when we divide by $s-t$ and pass to the limit, we arrive at

$$\dot{g}_t(z) = -g_t(z)\frac{\lambda(t) + g_t(z)}{\lambda(t) - g_t(z)}\partial_t(\log(a_1(t))), \tag{5.6}$$

provided that $a_1(t)$ is differentiable.

Remark 5.6. Quite often, Loewner equation is written as

$$\dot{g}_t(z) = g_t(z)\frac{1 + \kappa(t)g_t(z)}{1 - \kappa(t)g_t(z)},$$

where $\kappa(t) = 1/\lambda(t) = \bar{\lambda}(t)$.

Since there are very few conformal maps that can be written explicitly, there are very few instances where the Loewner evolution can be verified explicitly. Let us consider the simplest example, where the slit is growing along the real line. In this case, the map g_t is known explicitly (see Exercise 2.8 and formula (2.19)):

$$g_t(z) = \frac{\sqrt{x}(1+z) + \sqrt{(x-z)(1-xz)}}{\sqrt{x}(1+z) - \sqrt{(x-z)(1-xz)}} = \frac{(1+x)^2}{4x}z + \cdots,$$

where $x = 1-t$. The expansion gives the value of $a_1(t)$. The driving function is particularly simple: $\lambda(t) = g_t(x) = 1$. It is possible to differentiate the expression for g_t and with some algebraic manipulations verify that

g_t satisfies the Loewner equation, but even in this case, the brute force computation is rather long, messy, and not illuminating.

Theorem 5.4 provides us the basic result which allows one to describe a family of conformal maps generated by a curve growing inside a domain. Now, we can briefly present some alternative setups.

External Loewner Evolution. Let Ω be a simply connected domain of the form $\widehat{\mathbb{C}} \setminus K$ where K is a compact set containing at least two points. Let $\gamma(t) : [0,T] \to \Omega$ be a simple continuous curve which starts on the boundary, that is, $\gamma(0) \in K$, but the rest of the curve is inside Ω. We also allow the case $T = \infty$. Let $\Omega_t = \Omega \setminus \gamma([0,t]) = \widehat{\mathbb{C}} \setminus K_t$, $f_t : \mathbb{D}_- \to \Omega_t$ be the Riemann map such that $f_t(\infty) = \infty$ and $f_t'(\infty) > 0$. As always, we denote the inverse of f_t by g_t. We also assume that the curve γ is parametrized in such a way that

$$\log f_t'(\infty) = \operatorname{cap}(K_t) = \operatorname{cap}(K) + t.$$

We denote the preimage of $\gamma(t)$ under f_t by $\lambda(t)$.

We define $\widetilde{\Omega}_t$ to be the set of points z such that $1/z \in \Omega_t$. In this case, the family of maps $\tilde{f}_t(z) = 1/f_t(z^{-1})$ satisfies the assumptions of Theorem 5.4; hence they satisfy the Loewner differential equation (5.3) with the driving function $\tilde{\lambda}(t) = \lambda(t)^{-1}$. Differentiating $f_t(z) = 1/\tilde{f}_t(z^{-1})$ with respect to t, we get

$$\begin{aligned}\dot{f}_t(z) &= -\frac{1}{\tilde{f}_t^2(1/z)}\dot{\tilde{f}}_t(1/z) = -\frac{1}{\tilde{f}_t^2(1/z)}\left(-\frac{1}{z}\tilde{f}_t'(1/z)\frac{\tilde{\lambda}(t)+1/z}{\tilde{\lambda}(t)-1/z}\right)\\ &= \left(-\frac{1}{\tilde{f}_t^2(1/z)}\tilde{f}_t'(1/z)\left(-\frac{1}{z^2}\right)\right)z\frac{z+\lambda(t)}{z-\lambda(t)} = f_t'(z)z\frac{z+\lambda(t)}{z-\lambda(t)}.\end{aligned}$$

This proves that the maps f_t and g_t satisfy the same differential equations as in Theorem 5.4. The initial conditions are $f_0(z) = f(z)$ and $g_0(z) = g(z) = f^{-1}(z)$.

Whole-Plane Loewner Evolution. Let us consider an extreme version of the previous setup where the starting compact is a singleton $\{0\}$ (we can choose any other point, but choosing 0 simplifies some formulas). In this case, we cannot formally apply the previous approach since the initial map $f : \mathbb{D}_- \to \Omega$ cannot be defined. But our derivation of the Loewner differential equations depended only on the incremental maps $\phi_{s,t}$ which are well defined; this suggests that our case should not be intrinsically more difficult. It just requires a proper setup.

Let $\gamma : [-\infty, T] \to \widehat{\mathbb{C}}$ be a simple continuous curve which starts at the origin, that is, $\gamma(t) \to 0$ as $t \to -\infty$. We assume that the curve is parametrized by capacity: $\text{cap}(\gamma([-\infty, t])) = t$. Note that this normalization forces the time parameter to start from $-\infty$. Indeed, by continuity, diameter of $\gamma([-\infty, t])$ tends to 0 as $t \to -\infty$. By corollary of the Koebe distortion theorem (see Proposition 3.25), this implies that $\text{cap}(\gamma([-\infty, t]) \to -\infty$. For every $t > -\infty$ we can define f_t to be the Riemann map from $\mathbb{D}_-$ onto $\Omega_t = \widehat{\mathbb{C}} \setminus K_t = \widehat{\mathbb{C}} \setminus \gamma([-\infty, t])$. Let us fix some finite s and split the curve into two parts, $K = \gamma([-\infty, s])$ and the rest of the curve. This fits our previous case and we can see that the functions f_t for $t > s$ satisfy the Loewner differential equation. Since f_t is independent of the choice of $s < t$, this is true for all values of t.

For g_t, we can apply the result of Exercise 3.9 to claim that $e^t g_t(z) \to z$ as $t \to -\infty$ uniformly on compacts. There is no universal limit like this for f_t: the behaviour of rescaled f_t depends on the infinitesimal behaviour of γ near the origin.

Exercise 5.1. Let $\gamma(t) = 4e^t$ and $\Omega_t = \widehat{\mathbb{C}} \setminus \gamma([-\infty, t])$. Compute the corresponding maps f_t and g_t. Verify that $e^t g_t(z) \to z$.

Loewner evolution for single-slit maps. Let us recall that a single-slit map is a conformal map from the unit disc onto $\mathbb{C}$ with a slit along a curve going to infinity. This setup is in some sense the extreme version of the usual evolution where the starting domain is the entire complex plane. It is very similar to the whole-plane Loewner evolution; they, essentially, differ by a conjugation by inversion.

Let $\gamma : (-\infty, T]$ be a simple continuous curve such that $\gamma(t) \to \infty$ as $t \to -\infty$. We also allow $T = \infty$; in this case, we require $\gamma(t) \to 0$ as $t \to \infty$. We assume that $\gamma(t) \neq 0$ and that f_t is the Riemann map from $\mathbb{D}$ onto $\mathbb{C} \setminus \gamma((-\infty, t])$ with $f_t(0) = 0$. We also assume that the curve is parametrized by capacity, that is, $f_t'(0) = e^{-t}$. As always, we denote f_t^{-1} by g_t.

If we fix any s and define $\Omega = \mathbb{C} \setminus \gamma((-\infty, s])$, then the evolution of f_t and g_t for $t > s$ can be described by the usual Loewner evolution. This proves that f_t and g_t satisfy differential equations (5.3) and (5.2) for all $t > s$. Since s is arbitrary, this is true for all t.

By the same argument as in the whole-plane case, we can show that $e^{-t} g_t(z) \to z$ as $t \to -\infty$ uniformly on compact sets.

Exercise 5.2. Let γ be a curve growing from $+\infty$ along the real line. Find the capacity parametrization of this curve, the corresponding maps f_t and g_t and verify that f_t satisfy the Loewner differential equation. Show that $e^{-t}g_t(z) \to z$ as $t \to -\infty$ uniformly on compact sets.

Exercise 5.3. Let $\gamma : (-\infty, T] \to \mathbb{C}$ be a simple curve parametrized by capacity such that $\gamma(t) \to \infty$ as $t \to -\infty$. Let f_t be the corresponding map and let $\lambda(t)$ be the corresponding driving function of the single-slit Loewner evolution. Consider a rescaled and rotated curve $\tilde{\gamma} = c\gamma$, where $c = re^{i\theta}$. Show that $\tilde{\gamma}(s) = c\gamma(s + \log(r))$ is parametrized by capacity and the corresponding driving function is $\tilde{\lambda}(s) = \lambda(t)e^{i\theta}$. This means that the rotation of the curve leads to the rotation of the driving function and the (real) rescaling corresponds to the shift in time.

Remark 5.7. Since the whole-plane Loewner evolution is essentially the same as the single-slit one, the same holds for the whole-plane Loewner evolution.

5.1.2. *Chordal Loewner evolution*

The chordal version is very similar in spirit to the radial one; the main difference is that here we consider a curve growing from one point on the boundary of a domain towards another boundary point. In this case, there is no fixed 'target' point inside the domain; instead we have a fixed 'target' point on the boundary. This suggests that it is more convenient to use Riemann maps normalized on the boundary and the most natural domain for this is the upper half-plane.

To simplify notation, we formulate our results for curves in $\mathbb{H}$ growing from 0 towards infinity. We assume that $\gamma : [0, T] \to \mathbb{C}$ is a simple continuous curve such that $\gamma(0) = 0$ and the rest of the curve lies in $\mathbb{H}$. If $T = \infty$, then we assume that $\gamma(t) \to \infty$ as $t \to \infty$. We also assume that the curve parametrized in such a way that the half-plane capacity $\operatorname{hcap}(\gamma([0, t]))$ is a differentiable function of t. Since it is a strictly increasing function, it is differentiable almost everywhere, so the last assumption is not terribly restrictive.

Similar to the radial case, we define $\mathbb{H}_t = \mathbb{H} \setminus \gamma([0, t])$ and let $g_t : \mathbb{H}_t \to \mathbb{H}$ be the Riemann map with thermodynamic normalization. As before, by f_t, we denote g_t^{-1} and for $s < t$, we define $\phi_{s,t} = g_s(f_t)$. Finally, we denote the image of the tip of the curve by $u(t)$, namely $u(t) = g_t(\gamma(t))$ (see Figure 5.3).

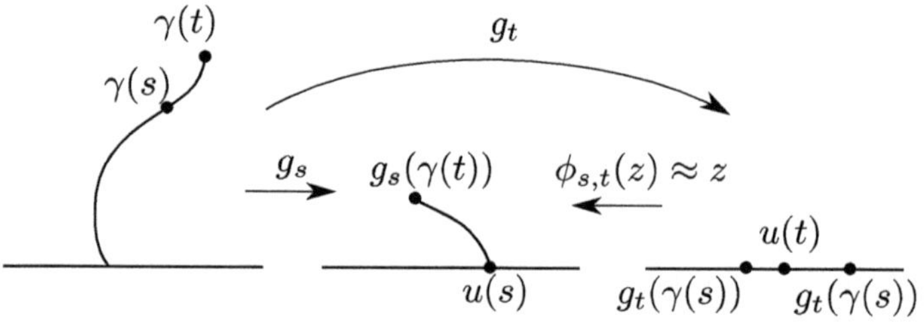

Fig. 5.3. There are two points denoted by $g_t(\gamma(s))$ that are mapped by $\phi_{s,t}$ to $u(s)$. Since $g_s(\gamma(t)) = \phi_{s,t}(u(t))$ is $c\sqrt{\varepsilon}$ close to $u(s)$ and $\phi_{s,t}(z)$ is uniformly close to z everywhere in the closed half-plane, $u(s)$ and $u(t)$ must also be $c\sqrt{\varepsilon}$ close.

Our goal is to derive differential equations satisfied by f_t and g_t. All proofs are very similar to the argument in the radial case, so we omit most of the details and only explain the main steps, paying attention to the parts where arguments are different.

The proof that $u(t)$ is continuous follows the same lines as the proof of continuity of $\lambda(t)$. The main difference is that in order to indicate that $\phi_{s,t}(z)$ converges to z, we show that the family $\phi_{s,t}(z) - z$ is normal.

Alternatively, we can use the results of mapping-out functions. Let s and t be so close that the diameter of $\gamma([s,t])$ is bounded by ε. By the half-plane version of Theorem 4.21, the diameter of $\gamma_{s,t} = g_s(\gamma([s,t]))$ is at most $c\sqrt{\varepsilon}$. By Lemma 3.33,

$$|\phi_{s,t}(z) - z| \le c\sqrt{\varepsilon}$$

everywhere in the closure of the upper half-plane. This immediately implies that $|u(t) - u(s)| \le c\sqrt{\varepsilon}$ (with some other, but still universal constant c). See Figure 5.3 for further details. Note that with this method, we do not have to separately consider limits as $s \to t$ and $t \to s$.

To derive the differential equation, we apply the half-plane version of Schwarz formula to $\phi_{s,t}(z) - z$ to write

$$\phi_{s,t}(z) - z = \frac{1}{\pi}\int \frac{1}{x-z}\operatorname{Im}\phi_{s,t}(x)\mathrm{d}x.$$

By construction, $g_s = \phi_{s,t}(g_t)$, so by plugging $g_t(z)$ instead of z in the above formula, we obtain

$$g_s(z) - g_t(z) = \frac{1}{\pi}\int \frac{1}{x-g_t(z)}\operatorname{Im}\phi_{s,t}(x)\mathrm{d}x.$$

Since $\operatorname{Im}\phi_{s,t}(x) = 0$ outside of a shrinking neighbourhood of $u(t)$, dividing by $s - t$ and passing to the limit as $s \nearrow t$, we obtain

$$\dot{g}_t(z) = \frac{1}{u(t) - g_t(z)} \lim \frac{1}{s-t}\frac{1}{\pi}\int \operatorname{Im}\phi_{s,t}(x)\mathrm{d}x.$$

By (3.20) and the additive property of capacity, the last integral is equal to $\operatorname{hcap}(\gamma([0,t])) - \operatorname{hcap}(\gamma([0,s]))$. This implies that the right-hand side is equal to

$$\frac{\partial_t \operatorname{hcap}(\gamma([0,t]))}{g_t(z) - u(t)}.$$

Now, we can formulate the main result.

Theorem 5.8. *Let γ be a curve in $\mathbb{H}$ satisfying all assumptions above and let f_t and g_t be the corresponding conformal maps and $u(t) = g_t(\gamma(t))$. Then g_t satisfies an ordinary differential equation*

$$\dot{g}_t(z) = \frac{\partial_t \operatorname{hcap}(\gamma([0,t]))}{g_t(z) - u(t)}, \quad g_0(z) = z, \tag{5.7}$$

and f_t satisfies

$$\dot{f}_t(z) = -f_t'(z)\frac{\partial_t \operatorname{hcap}(\gamma([0,t]))}{z - u(t)}, \quad f_0(z) = z. \tag{5.8}$$

These differential equations are called *chordal Loewner differential equations* or *chordal Loewner evolutions*, and the function $u(t)$ is called the *driving function.*

Proof of Theorem 5.8. We have already sketched the proof of the first part. To obtain the second, we can use the inverse function theorem to show that f_t is differentiable with respect to t. After this, we write $f_t(g_t(z)) = z$, differentiate it with respect to t and plug in the equation for $\dot{g}_t$. All technical details are virtually the same as in the radial case and are left to the reader. □

In this theorem, we assumed that the capacity is differentiable. This seems to be very restrictive but, in some sense, it is not. Half-plane capacity is a strictly increasing function; hence by an appropriate change of variables, we can reparametrize any Jordan curve γ in such a way that its capacity becomes any other strictly increasing function. In particular, we can make it differentiable.

There is a canonical parametrization such that $\operatorname{hcap}(\gamma([0,t])) = 2t$. This is called *parametrization by capacity.* Since capacity is additive, it is natural to make it a linear function of time. Somewhat arbitrary, constant 2 initially appeared from a connection between chordal and radial evolutions. Nowadays, it is used mostly for historical and consistency reasons.

Exercise 5.4. Let γ be a straight vertical interval in $\mathbb{H}$ growing from the origin. Find its capacity parametrization, corresponding conformal maps, and verify that they satisfy the Loewner differential equation.

The main goal of this section was to derive Loewner differential equations, so we did it in the simplest case of properly parametrized Jordan curves. In fact, the same results could be obtained in a more general setting of compact hulls having the *local growth* property, namely, if we consider a growing family of hulls K_t such that $K_{t,t+\varepsilon} = g_{K_t}(K_{t+\varepsilon} \setminus K_t)$ shrinks, that is, $\mathrm{rad}(K_{t,t+\varepsilon}) \to 0$ as $\varepsilon \to 0$ locally uniformly in t, then the family of maps $g_t = g_{K_t}$ satisfies the Loewner differential equation with the driving function $u(t) = \bigcap_\varepsilon K_{t,t+\varepsilon}$. The complete proof of this statement is very similar to the proof of Theorem 5.8.

We would like to conclude with several remarks. First, our derivation of (5.7) is based on the analysis of $\phi_{s,t}$, so essentially the same argument works for curves growing from a boundary point in an arbitrary domain. The only important part is that the maps should be normalized at some other boundary point. The only reason why we stated the above theorem for curves in $\mathbb{H}$ is to avoid a discussion of what capacity is in other domains.

Although chordal and radial Loewner evolutions are very similar in nature, there are also some very important differences. Some of them are more subtle and become evident later, but even now we can mention two of them. In the radial case, when we parametrize by capacity, it is possible to pass to the limit as time starts from $-\infty$; the result is the whole-plane Loewner evolution where the curve grows from a single point. There is no equivalent construction in the chordal case. But chordal evolution has an additional nice property in that it behaves particularly simply when we scale the curve (see also Exercise 5.3 for scaling of the whole-plane and single-slit Loewner evolutions). To be more precise, let $\lambda > 0$ be a scaling factor and consider the curve γ rescaled by λ. We assume that γ is parametrized by capacity. Since we know how capacity scales, this allows us to parametrize the scaled curve by capacity. The exact parametrization is $\tilde{\gamma}(t) = \lambda\gamma(t/\lambda^2)$. We denote the corresponding family of maps by $\tilde{g}_t$. These maps are also easy to find. A direct computation shows that

$$\tilde{g}_t(z) = \lambda g_{t/\lambda^2}(z/\lambda) \tag{5.9}$$

is a conformal map from the complement of $\tilde{\gamma}([0,t])$ onto $\mathbb{H}$ and has the thermodynamic normalization (see also Proposition 3.26). It is immediately clear that

$$\tilde{u}(t) = \tilde{g}_t(\tilde{\gamma}(t) = \lambda g_{t/\lambda^2}(\gamma(t/\lambda^2)) = \lambda u(t/\lambda^2). \tag{5.10}$$

Differentiating (5.9) with respect to t or directly applying Theorem 5.8, we have

$$\dot{\tilde{g}}_t(z) = \frac{2}{\tilde{g}_t(z) - \tilde{u}_t}. \tag{5.11}$$

If the curve γ is invariant under scaling, which happens if and only if it is a straight interval from the origin, then the driving function $u(t)$ should be invariant under rescaling given by (5.10). This suggests that the driving function for the straight interval should be of the form $c\sqrt{t}$, where c is related to the direction of the interval. This is indeed true and can be verified by a direct computation.

Exercise 5.5. Let γ be a straight interval in $\mathbb{H}$ from the origin forming angle $\alpha\pi$ with the positive real line. Parametrize it by capacity, find the corresponding maps f_t, and verify that they satisfy the Loewner differential equation (5.8).

Remark 5.9. Functions of the form $c\sqrt{t}$ are the only continuous functions that are invariant under scaling given by (5.10). Everyone familiar with probability immediately recognizes this scaling: it is the Brownian motion scaling, that is, the Brownian motion is invariant under this scaling. This fact plays a crucial role in the theory of Schramm–Loewner evolution, which is an extremely important probabilistic counterpart of the Loewner evolution.

Exercise 5.6. Let $\gamma(t)$ be a curve in $\mathbb{H}$ parametrized by capacity. Define $\tilde{\gamma}(t) = \gamma(t) + x$, where x is real. Let $u(t)$ be the driving function of γ. Show that $\tilde{u}(t) = u(t) + x$ is the driving function of $\tilde{\gamma}$.

Exercise 5.7. Let γ be a curve parametrized by capacity and let $u(t)$ be its driving function. We denote the symmetric image of γ with respect to the imaginary axis by $\tilde{\gamma}(t)$, that is, $\operatorname{Im}\gamma(t) = \operatorname{Im}\tilde{\gamma}(t)$ and $\operatorname{Re}\gamma(t) = -\operatorname{Re}\tilde{\gamma}(t)$. Show that the driving function for $\tilde{\gamma}$ is $\tilde{u}(t) = -u(t)$.

5.1.3. *Applications*

Before studying the Loewner evolution in more detail and grater generality, we would like to present one of its first applications. Many applications of the Loewner differential equation arise from the fact that single-slit maps are dense in S (see Theorem 3.18).

In particular, this means that if we want to optimize any functional which is continuous with respect to the locally uniform convergence, then it is enough to optimize it over slit maps. Since the Loewner evolution is a very powerful tool to study slit maps, it also helps obtain the results of *all* functions from S.

We start by studying the first two non-trivial coefficients of functions from class S. These results were the first applications of the Loewner evolution and were the main reason for introducing the Loewner evolution.

Let us recall that the functions from class S are the univalent functions in $\mathbb{D}$ of the form $f(z) = z + \sum a_n z^n$ and that the Bieberbach conjecture states that $|a_n| \leq n$. In this section, we give an alternative proof of the Bieberbach conjecture for $n = 2$ and a proof for $n = 3$.

By the Cauchy formula, if $f_n \to f$ uniformly on compacts, then $f_n' \to f'$. Since Taylor coefficients can be written in terms of derivatives, this means that all functionals $f \mapsto a_n$ are continuous on S. Hence, to prove that $|a_n| \leq n$, it is sufficient to prove the inequality for single-slit maps.

Let f be a single-slit map and let γ be the corresponding slit. We parametrize γ in such a way that $\gamma((-\infty, 0]) \to \mathbb{C}$ and $\gamma(t) \to \infty$ as $t \to -\infty$. Let f_t be the corresponding Riemann maps

$$f_t : \mathbb{D} \to \mathbb{C} \setminus \gamma((-\infty, t]).$$

We assume that $f_t'(0) = e^{-t}$ and we denote f_t^{-1} by g_t. Note that the time runs up to $t = 0$, which exactly corresponds to $f = f_0 \in S$.

Let us consider a family of maps $h_t(z) = g_t(f(z))$. Since g_t satisfies the Loewner differential equation (5.2), it is easy to see that h_t is also a solution to this equation. Indeed, differentiating with respect to t, we obtain

$$\dot{h}_t = \dot{g}_t(f) = g_t(f)\frac{\lambda(t) + g_t(f)}{\lambda(t) - g_t(f)} = h_t\frac{\lambda(t) + h_t}{\lambda(t) - h_t}.$$

Let us consider the family of maps $e^{-t}h_t$. Since $f'(0) = 1$ and $g_t'(0) = e^t$, it is clear that $e^{-t}h_t \in S$. From the definition of h_t, it is immediately clear that $e^{-0}h_0(z) = z$. But, by Exercise 3.9, maps $e^{-t}g_t(w) \to w$ uniformly on compacts and $e^{-t}h_t$ converges to f as $t \to -\infty$. In some sense, this means that maps $e^{-t}h_t$ interpolate between the identity and f inside S.

Since the function h_t is analytic in z and differentiable with respect to t, its Taylor coefficients with respect to z are also differentiable with respect to t. We introduce notation

$$h_t(z) = e^t(z + a_2(t)z^2 + a_3(t)z^3 + \cdots).$$

Differentiating the series with respect to t, we obtain

$$e^t(z + (a_2(t) + \dot{a}_2(t))z^2 + (a_3(t) + \dot{a}_3(t))z^3 + \cdots).$$

But by the Loewner differential equation,

$$\begin{aligned}\dot{h}_t(z) &= h_t(z)\frac{\lambda(t)+h_t(z)}{\lambda(t)-h_t(z)}\\ &= e^t\left(z+\left(a_2(t)+\frac{2e^t}{\lambda(t)}\right)z^2\right.\\ &\quad\left.+\left(a_3(t)+\frac{2e^{2t}}{\lambda(t)^2}+\frac{4a_2(t)e^t}{\lambda(t)}\right)z^3+\cdots\right).\end{aligned}$$

Equating the coefficients in front of z^2 and z^3, we obtain two differential equations:

$$\dot{a}_2(t) = \frac{2e^t}{\lambda(t)} \tag{5.12}$$

and

$$\dot{a}_3(t) = \frac{2e^{2t}}{\lambda(t)^2} + \frac{4a_2(t)e^t}{\lambda(t)}. \tag{5.13}$$

We also know that $a_k(0) = 0$ for all $k \geq 2$ and $a_k(t) \to a_k$ as $t \to -\infty$, where a_k are the Taylor coefficients of f.

We start by analyzing the first differential equation. Integrating with respect to t, we have

$$a_2 = a_2(-\infty) - a_2(0) = -\int_{-\infty}^{0} \dot{a}_2(s)\mathrm{d}s = -2\int_{-\infty}^{0} \frac{e^s}{\lambda(s)}\mathrm{d}s.$$

Since $|\lambda(t)| = 1$, we can estimate

$$|a_2| \leq 2\int_{-\infty}^{0} e^s \mathrm{d}s = 2.$$

Moreover, we can see that equality occurs if and only if $\lambda(t)$ is constant. We already know that $\lambda(t) \equiv 1$ corresponds to the Koebe function. It is easy to believe and straightforward to verify that $\lambda(t) \equiv e^{i\theta}$ corresponds to a rotation of the Koebe function. This gives an alternative proof of Theorem 3.3.

To analyze the equation for the third coefficient, we rewrite it using the equation for the second coefficient (5.12):

$$\dot{a}_3(t) = 2e^{2t}\lambda^{-2}(t) + 2\dot{a}_2(t)a_2(t).$$

Integrating and using an integral formula for a_2, we have

$$\begin{aligned}
a_3 &= -\int_{-\infty}^{0} \dot{a}_3(s)\mathrm{d}s = -2\int_{-\infty}^{0} e^{2s}\lambda^{-2}(s)\mathrm{d}s - \int_{-\infty}^{0} 2\dot{a}_2(s)a_2(s)\mathrm{d}s \\
&= -2\int_{-\infty}^{0} e^{2s}\lambda^{-2}(s)\mathrm{d}s + a_2^2 \\
&= -2\int_{-\infty}^{0} e^{2s}\lambda^{-2}(s)\mathrm{d}s + 4\left(\int_{-\infty}^{0} e^{s}\lambda^{-1}(s)\mathrm{d}s\right)^2,
\end{aligned}$$

where, in the last line, we used the explicit integral representation for a_2 that we obtained before. Next, we write $\lambda(s) = e^{i\theta(s)}$ and take the real part of the above equation:

$$\begin{aligned}
\mathrm{Re}(a_3) = -2\int_{-\infty}^{0} e^{2s}\cos(2\theta(s))\mathrm{d}s + 4\left(\int_{-\infty}^{0} e^{s}\cos(\theta(s))\mathrm{d}s\right)^2 \\
- 4\left(\int_{-\infty}^{0} e^{s}\sin(\theta(s))\mathrm{d}s\right)^2.
\end{aligned}$$

Removing the last term, using the double angle formula and the Cauchy–Schwarz inequality, we have

$$\begin{aligned}
\mathrm{Re}(a_3) &\le -2\left(2\int_{-\infty}^{0} e^{2s}\cos^2(\theta(s))\mathrm{d}s - \int_{-\infty}^{0} e^{2s}\mathrm{d}s\right) \\
&\quad + 4\int_{-\infty}^{0} e^{s}\mathrm{d}s \int_{-\infty}^{0} e^{s}\cos^2(\theta(s))\mathrm{d}s \\
&= 1 + 4\int_{-\infty}^{0} \left(e^{s} - e^{2s}\right)\cos^2(\theta(s))\mathrm{d}s \\
&\le 1 + 4\int_{-\infty}^{0} \left(e^{s} - e^{2s}\right)\mathrm{d}s = 3.
\end{aligned}$$

Finally, let us write the third coefficient of f as $|a_3|e^{i\theta}$ and consider the function $\tilde{f}(z) = e^{i\theta/2}f(e^{-i\theta/2}z)$ which is also from class S. Direct computation shows that its third coefficient is $e^{-i\theta}|a_3|e^{i\theta} = |a_3|$. Applying the previous argument to $\tilde{a}_3$, the third coefficient of $\tilde{f}$, we arrive at

$$|a_3| = \tilde{a}_3 = \mathrm{Re}(\tilde{a}_3) \le 3.$$

This completes the proof of the Bieberbach conjecture for $n = 3$.

Remark 5.10. Using the same method, one can write similar, but increasingly more complicated, differential equations on a_n in terms of a_k for $k \le n$.

Direct analysis of these equations becomes exceedingly difficult. Nevertheless, the Loewner evolution played a crucial role in all proofs of the partial cases of the Bieberbach conjecture and in its final solution by de Branges (1985).

There are many other applications of the Loewner techniques; in particular, the sharp form of Theorem 3.10 is obtained using the Loewner evolution. To give some idea of how these methods work, we present the proof of one of the Koebe distortion formulas (3.10). We start by proving the inequality due to Grunsky.

Theorem 5.11. *For f from the class S and $z \in \mathbb{D}$, the following inequality holds:*

$$\left|\log\left(z\frac{f'(z)}{f(z)}\right)\right| \le \log\frac{1+|z|}{1-|z|}. \tag{5.14}$$

Before proving this, theorem we note that by taking the real part of this inequality, we obtain

$$-\log\frac{1+|z|}{1-|z|} \le \log\left|z\frac{f'(z)}{f(z)}\right| \le \log\frac{1+|z|}{1-|z|},$$

which after exponentiation becomes (3.10):

$$\frac{1-|z|}{1+|z|} \le \left|z\frac{f'(z)}{f(z)}\right| \le \frac{1+|z|}{1-|z|}.$$

Remark 5.12. In Exercise 3.8, we derived this inequality from the Growth theorem. In a similar way, the Growth theorem can be derived from this inequality. Together, these two estimates imply the Distortion theorem. This shows that all these results could be derived using the Loewner evolution.

Proof of Theorem 5.11. As above, for a slit map f, we construct a family of maps h_t which solves the Loewner equation with the driving function $\lambda(t)$, $h_0(z) = z$, and $e^{-t}h_t \to f$ as $t \to -\infty$. From the Loewner equation for h_t, we obtain

$$\partial_t \log h_t = \frac{\lambda(t)+h_t}{\lambda(t)-h_t}.$$

Taking the real part, we derive

$$\partial_t|h_t| = |h_t|\,\mathrm{Re}\left(\frac{\lambda(t)+h_t}{\lambda(t)-h_t}\right) = |h_t|\frac{1-|h_t|^2}{|\lambda(t)-h_t|^2}. \tag{5.15}$$

Assuming that h_t' is differentiable with respect to t (this will follow from an argument in the proof of Theorem 5.16) and using the Loewner differential equation, we have

$$\partial_t \log\left(z\frac{h_t'(z)}{h_t(z)}\right) = h_t\frac{2\lambda(t)}{(\lambda(t)-h_t)^2}.$$

Therefore, integrating from $-\infty$ to 0, we obtain

$$\int_{-\infty}^{0} h_t\frac{2\lambda(t)}{(\lambda(t)-h_t(z))^2}\,\mathrm{d}t = \log\left(z\frac{h_0'(z)}{h_0(z)}\right) - \lim_{t\to-\infty}\log\left(z\frac{h_t'(z)}{h_t(z)}\right)$$
$$= -\log\left(z\frac{f'(z)}{f(z)}\right).$$

By the triangle inequality and (5.15),

$$\left|\log\left(z\frac{f'(z)}{f(z)}\right)\right| \leq \int_{-\infty}^{0}\frac{2|h_t(z)|}{|\lambda(t)-h_t(z)|^2}\,\mathrm{d}t$$
$$= \int_{-\infty}^{0}\frac{2}{1-|h_t(z)|^2}\partial_t|h_t(z)|\mathrm{d}t$$
$$= \int_{0}^{|z|}\frac{2}{1-s^2}\,\mathrm{d}s = \log\frac{1+|z|}{1-|z|},$$

where we used the change of variables $s = |h_t(z)|$, the fact that $|h_0(z)| = |z|$ and that $|h_t(z)| \to 0$ as $t \to -\infty$. This completes the proof of (5.14) in the case when f is a slit map. Since the slit maps are dense in S and for a fixed $z \in \mathbb{D}$ the functional

$$f \mapsto \left|\log\left(z\frac{f'(z)}{f(z)}\right)\right|$$

is continuous with respect to the local uniform convergence, the same inequality holds for all $f \in S$. □

Many other examples showing the power of Loewner's method can be found in Duren (1983) and Pommerenke (1975).

5.2. Loewner Chains

In the previous section, we studied families of simply connected domains obtained by growing a simple slit. In this section, we study more general families of decreasing domains and will show that they also satisfy a differential equation similar to the one we discussed above.

Instead of domains obtained by removing a growing curve from an initial domain, we consider a rather general family of shrinking domains. The formal setup is given by the following definition.

Definition 5.13. A continuous function $f : \mathbb{D} \times [0, T) \to \mathbb{C}$ is called a *Loewner chain* if the following three properties hold:

(1) For every t, the map $f_t(z) = f(z, t)$ is analytic and univalent.
(2) For every t, we have $f_t(0) = 0$ and $f_t'(z) = e^{-t}$. Equivalently,

$$f_t(z) = e^{-t}z + a_2(t)z^2 + a_3(t)z^3 + \cdots .$$

(3) Domains $\Omega_t = f_t(\mathbb{D})$ are shrinking, namely $\Omega_t \subset \Omega_s$ if $s \leq t$.

The condition $f_t'(z) = e^{-t}$ is not crucial to the theory of Loewner chains, but gives a convenient time parametrization.

Since all maps f_t are univalent and preserve the origin, the last condition could be replaced by $f_t \prec f_s$ for $s \leq t$ (see Section 3.3).

Remark 5.14. As in the previous definition of the Loewner evolution, we use a parametrization different from the one found in many classical books. In the classical treatment, domains are increasing to the entire complex plane.

Our definition is given in terms of conformal maps. Sometimes, it is more convenient to formulate it in terms of the family of domains Ω_t. It is easy to check that if f_t is a Loewner chain then $\Omega_t \to \Omega_s$ in the sense of kernels as $t \to s$. The converse is almost true up to a rescaling and a reparametrization.

Lemma 5.15. *Let Ω_t be a family of simply connected domains such that $0 \in \Omega_t$, $\Omega_{t_2} \subsetneq \Omega_{t_1}$ for $t_1 < t_2$. Let $h_t(z)$ be the family of univalent maps from $\mathbb{D}$ onto Ω_t such that $h_t(0) = 0$ and $h_t'(0) = a_1(t) > 0$. Then the function $a_1(t)$ is strictly decreasing and the family of maps*

$$f_s(z) = \frac{1}{a_1(0)} h_t(z)$$

is a Loewner chain. Here, $t = t(s)$ is the inverse function of

$$s(t) = \log\left(\frac{a_1(0)}{a_1(t)}\right).$$

The change of time $s(t)$ is to ensure that the new derivative at the origin behaves as e^{-s}. The rescaling by $a_1(0)$ is also necessary because the initial map in the Loewner chain is assumed to be from S.

The proof of this lemma is essentially the same as in the slit case and is left to the reader.

Our main goal is to show that a Loewner chain $f_t(z)$ and the corresponding family of inverse maps $g_t(z)$ satisfy differential equations similar to equations (5.3) and (5.2) that we have already derived for slit maps.

Theorem 5.16. *Let f_t be a Loewner chain and g_t be the family of inverse maps $g_t = f_t^{-1}$. Then there exists a family of maps $p_t(z)$ such that for almost all t*

$$\dot{f}_t(z) = -z f_t'(z) p_t(z), \tag{5.16}$$

$$\dot{g}_t(z) = g_t(z) p_t(g_t(z)). \tag{5.17}$$

The family of maps p_t is measurable in t and for each t it belongs to the Carathéodory class $\mathcal{P}$.

These equations are called the Loewner or Loewner–Kufarev differential equations and are a direct analogue of the Loewner equations for slit maps. Indeed, if we consider

$$p_t(z) = \frac{\lambda(t) + z}{\lambda(t) + z},$$

which is obviously from $\mathcal{P}$, then the above equations become exactly the slit-map Loewner equations (5.3) and (5.2).

The similarity becomes even more evident if we represent functions p_t using the Herglotz theorem (Theorem 3.24). Using this theorem, the Loewner equations become

$$\dot{f}_t(z) = -z f_t'(z) \int_0^{2\pi} \frac{e^{i\theta} + z}{e^{i\theta} - z} \mu_t(\mathrm{d}\theta) \tag{5.18}$$

and

$$\dot{g}_t(z) = g_t(z) \int_0^{2\pi} \frac{e^{i\theta} + g_t(z)}{e^{i\theta} - g_t(z)} \mu_t(\mathrm{d}\theta), \tag{5.19}$$

where μ_t is a family of probability measures. In this form, it is immediately clear that the slit-map equation corresponds to the case when μ_t are δ-measures at $\arg(\lambda(t))$. Sometimes, it is more convenient to identify μ_t with a probability measure on the unit circle $\partial\mathbb{D}$ and write the equations as

$$\dot{f}_t(z) = -z f_t'(z) \int_{\partial\mathbb{D}} \frac{\zeta + z}{\zeta - z} \mu_t(\mathrm{d}\zeta),$$

$$\dot{g}_t(z) = g_t(z) \int_{\partial\mathbb{D}} \frac{\zeta + g_t(z)}{\zeta - g_t(z)} \mu_t(\mathrm{d}\zeta).$$

Similar to the slit-map case, we say that μ is the *driving measure*.

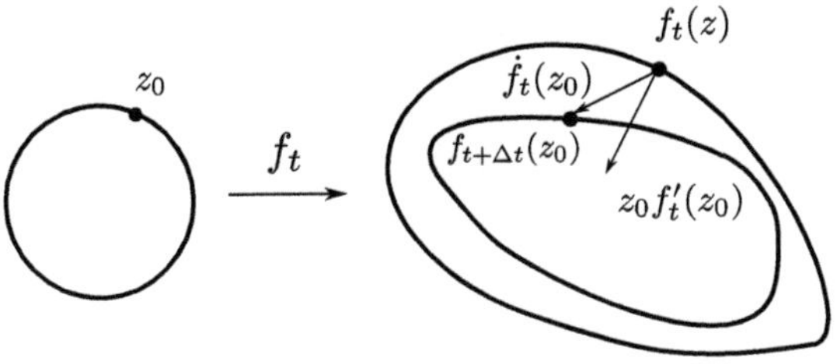

Fig. 5.4. Points under the Loewner evolution move 'inwards'.

At the moment, equations (5.16) and (5.17) might look mysterious. In particular, it is not clear where the condition that $p \in \mathcal{P}$ is coming from. It is possible to give these equations a simple geometric explanation.

Let us consider the first equation (5.16). We could interpret $f_t(z)$ as a trajectory of a *fixed* point z as the time t changes. For some t and $r < 1$, we can consider a closed curve $f_t(re^{i\theta})$ with $\theta \in [0, 2\pi)$. Let z_0 be some point on the circle of radius r. The inward normal vector to this curve at the point $f_t(z_0)$ is given by $z_0 f_t'(z_0)$. Since $\operatorname{Re} p_t > 0$, $\arg(p_t) \in (-\pi/2, \pi/2)$ and equation (5.16) implies that the direction of the velocity vector $\dot{f}_t(z_0)$ differs from the direction of the inward normal by at most $\pi/2$. This means that the flow $f_t(z)$ always goes 'inwards', which corresponds to the fact that the domains Ω_t are shrinking (see Figure 5.4). In some sense, the distribution of the measure μ_t describes how fast different parts of the boundary move inwards.

Another way of approaching these equations is in terms of driving measures. By the Herglotz theorem, the factor p in equations (5.16) and (5.17) can be written in the integral form as in equations (5.18) and (5.19). These integrals are averaging the Schwarz kernel with respect to μ. This kernel appears in the slit-map version of the Loewner evolution. This suggests that one can assume that these equations describe the growing infinitesimal slits everywhere on the boundary and the measure μ_t controls how fast they are growing at different parts of the boundary.

Unlike the slit-map case, the standard proof of Theorem 5.16 is based on the analysis of maps f_t rather than g_t. This conveniently provides a different perspective on the nature of the Loewner equations.

Proof of Theorem 5.16. We start by introducing the *transition function* of the Loewner chain, defined by $\phi_{s,t} = g_s(f_t)$ for $s \le t$. This is well defined since $\Omega_t \subset \Omega_s$. From the definition of the Loewner chain, we have that $\phi_{s,t}$ is univalent in the unit disc, $|\phi_{s,t}| \le 1$ and $\phi_{s,t}'(0) = e^{s-t}$. It is also clear that $f_t(z) = f_s(\phi_{s,t}(z))$.

Next, we introduce the function

$$p_{s,t}(z) = \frac{1+e^{s-t}}{1-e^{s-t}} \frac{z-\phi_{s,t}(z)}{z+\phi_{s,t}(z)} = \frac{1+e^{s-t}}{1-e^{s-t}} \frac{1-\phi_{s,t}(z)/z}{1+\phi_{s,t}(z)/z}.$$

From normalization of $\phi_{s,t}$ we can see that $p_{s,t}$ has a removable singularity at $z=0$. We remove this singularity by defining $p_{s,t}(0)=1$. By the Schwarz lemma (Theorem 2.2), $|\phi_{s,t}(z)| \le |z|$; this implies that the second factor in the definition of $p_{s,t}$ has a positive real part. The first factor is positive for $s<t$. This implies that for all $s<t$, the function $p_{s,t}$ belongs to the Carathéodory class $\mathcal{P}$ and we can apply a universal upper bound (3.15)

$$|p_{s,t}(z)| \le \frac{1+|z|}{1-|z|}.$$

This estimate allows us to control how fast $\phi_{s,t}$ converges to z as $s-t \to 0$. Indeed, the definition of $p_{s,t}$ can be rewritten as

$$z-\phi_{s,t}(z) = \frac{1-e^{s-t}}{1+e^{s-t}} p_{s,t}(z)(z+\phi_{s,t}(z)).$$

Using the upper bound for $p_{s,t}$ and the fact that $|\phi_{s,t}(z)| \le |z|$, we get

$$|z-\phi_{s,t}(z)| \le (1-e^{s-t})\frac{1+|z|}{1-|z|}(2|z|). \tag{5.20}$$

In particular, this implies that $\phi_{s,t}$ converges to z uniformly on compact sets.

This inequality allows us to control how f_t depends on t. Using the definition of $\phi_{s,t}$, we can write

$$\begin{aligned} |f_t(z)-f_s(z)| &= |f_s(\phi_{s,t}(z))-f_s(z)| = \left| \int_{\phi_{s,t}(z)}^{z} f_s'(\zeta)\mathrm{d}\zeta \right| \\ &\le |z-\phi_{s,t}(z)| \sup_{\zeta\in[z,\phi_{s,t}(z)]} |f_s'(\zeta)| \\ &\le (1-e^{s-t})\frac{1+|z|}{1-|z|} 2|z|e^{-s}\frac{1+|z|}{(1-|z|)^3} \\ &\le \frac{8|z|}{(1-|z|)^4}(e^{-s}-e^{-t}). \end{aligned} \tag{5.21}$$

The second inequality follows from the inequality (5.20) and the Distortion theorem (Theorem 3.9) which can be applied since $e^s f_s$ belongs to the class S, and all points ζ on the interval $[\phi_{s,t}(z), z]$ have modulus at most $|z|$.

The inequality (5.21) implies that f_t locally uniformly converges to f_s as $t \to s$, which in turn implies that $f_t'(z) \to f_s'(z)$ (by symmetry, a similar statement holds for $s \to t$ or even for $t - s \to 0$). It also implies that $f_t(z)$ is absolutely continuous with respect to t for every $z \in \mathbb{D}$. Since absolutely continuous functions are differentiable almost everywhere, the time derivative $\dot{f}_t(z)$ exists for almost all t. The set of exceptional times might depend on z and it is *a priori* not clear that $\dot{f}_t(z)$ exists for all z for a.e. t. We prove this by Theorem 2.11. Since the countable union of null sets is a null set, there is a set $\mathcal{T}$ of full measure such that on this set of times, $\dot{f}_t(z)$ exists for all z of the form $1/n$ with $n = 2, 3, \ldots$. This set has a point of accumulation inside $\mathbb{D}$; hence, by Vitali's theorem, $\dot{f}_t(z)$ exists for all $z \in \mathbb{D}$ and $t \in \mathcal{T}$.

Finally, we write

$$\frac{f_t(z) - f_s(z)}{t - s} = -\frac{1 - e^{s-t}}{t - s} \frac{z + \phi_{s,t}(z)}{1 + e^{s-t}} \frac{f_s(\phi_{s,t}(z)) - f_s(z)}{\phi_{s,t}(z) - z} p_{s,t}(z). \quad (5.22)$$

This identity is easy to verify by plugging in the definition of $p_{s,t}$ into the right-hand side and observing that most terms will cancel out.

Now, we would like to pass to the limit as $t \to s$. On the right-hand side, the first factor trivially converges to 1 and the second converges to z. The third factor for $s \in \mathcal{T}$ can be written as

$$\int_0^1 f_s'(\alpha z + (1 - \alpha)\phi_{s,t}(z))\mathrm{d}\alpha.$$

Since $\phi_{s,t}(z) \to z$ locally uniformly, the integral converges to $f_s'(z)$. Now, for $s \in \mathcal{T}$, the left-hand side of (5.22) converges to $\dot{f}_s$ and the first three terms on the right-hand side converge to $zf_s'(z)$. This implies that the last factor $p_{s,t}(z)$ also converges locally uniformly to some function $p_s(z)$ which obviously also belongs to the class $\mathcal{P}$. Combining all these results, we arrive at

$$\dot{f}_s(z) = -zf_s'(z)p_t(z), \quad s \in \mathcal{T},$$

which is exactly equation (5.16).

The equation on g_t is obtained from the equation on f in exactly the same way as we have done it before in the case of slit maps. □

This completes the derivation of the Loewner equations in the standard setup. Essentially, the same argument as in the slit-map case shows that these equations are also valid for the external and whole-plane versions of the Loewner chain.

The Loewner equations (5.18) and (5.19) involve the integral of the Schwarz kernel in the unit disc with respect to a measure, which should not be surprising since our proof uses the Herglotz representation theorem, which is based on the Schwarz kernel.

If we consider a family of continuously shrinking domains Ω_t without assumptions that they are parametrized by conformal radius, then the same argument will work, but measures μ_t are no longer probability measures. The precise relation between the mass of μ_t and time parametrization will be discussed in Section 5.3.

In the chordal case, the situation is almost the same. Instead of the Schwarz formula in the unit disc, we should use its half-plane version given in Proposition 2.40. Assuming the domains $\Omega_t = \mathbb{H} \setminus K_t$, where K_t is a continuous family of growing half-plane hulls and that they are normalized by the half-plane capacity, that is, $\operatorname{hcap} K_t = 2t$, essentially, the same argument implies that for half-plane Loewner chains, we have the following Loewner equations:

$$\dot{f}_t(z) = -f_t'(z) \int_{\mathbb{R}} \frac{\mu_t(\mathrm{d}x)}{z - x}, \tag{5.23}$$

$$\dot{g}_t(z) = \int_{\mathbb{R}} \frac{\mu_t(\mathrm{d}x)}{g_t(z) - x}, \tag{5.24}$$

where μ_t is a family of compactly supported measures on $\mathbb{R}$ such that $\mu_t(\mathbb{R}) = 2$.

Exercise 5.8. Let f_t be a radial Loewner chain and let $p_t(z)$ be the corresponding function from the Carathéodory class. For an integer n, we consider a symmetrization of f_t, namely, the function $\tilde{f}_t(z) = (f_{nt}(z^n))^{1/n}$. Show that $\tilde{f}_t$ form a Loewner chain which satisfies the Loewner equation with $\tilde{p}_t(z) = p_{nt}(z^n)$.

5.3. Solving the Loewner Evolution

In the two previous sections, we have shown that under rather mild assumptions, families of continuously shrinking or growing domains can be described by simple differential equations that are driven by a family of measures.

What makes the Loewner evolution especially useful is that this process can be reversed. Given a family of *driving measures* satisfying very mild assumptions, one can solve Loewner differential equations and it turns out that the solution is a family of conformal maps which corresponds to a Loewner chain. This means that these equations provide a correspondence between families of shrinking/growing domains and families of measures.

As before, we start by analyzing the radial Loewner evolution. In most of the theorems before, it was not that important whether we started with f_t or g_t: sometimes one was a bit more convenient than the other, but the difference was not crucial. When it comes to solving the equations, the difference is huge. If we fix z, and $g_0(z)$, then the equation for $g_t(z)$ becomes a first-order nonlinear ordinary differential equation. General theory of these equations is very well developed and under very mild assumptions, the solutions to these equations exist and are unique. So, essentially, we just have to study the properties of these solutions. But the equation for $f_t(z)$ is a nonlinear partial differential equation, and even showing that they have a (unique) solution is significantly more involved. For these reasons, we start by analyzing the equation for g_t.

Theorem 5.17. *Let μ_t, $t \in [0, \infty)$, be a family of non-negative Borel measures on $[0, 2\pi)$ which is locally uniformly bounded, that is,*

$$\sup_{0 \le t \le T} \mu_t([0, 2\pi)) < \infty, \quad \textit{for all } T > 0.$$

We also assume that μ_t is continuous with respect to t in the weak topology. For every $z \in \mathbb{D}$, let $g_t(z)$ be the solution of the Loewner differential equation

$$\dot{g}_t(z) = g_t(z) \int_0^{2\pi} \frac{e^{i\theta} + g_t(z)}{e^{i\theta} - g_t(z)} \mu_t(\mathrm{d}\theta)$$

with the initial condition $g_0(z) = z$. We denote the maximal time interval of existence of the solution of the initial value problem inside the unit disc by $[0, T(z))$. We denote the set of points such that $T(z) \le t$ and by $\Omega_t = \{z \in \mathbb{D} : T(z) > t\}$ the set of the initial conditions for which the solution exists up to the time t by K_t. Then $g_t(z)$ is the unique conformal transformation from Ω_t onto the unit disc normalized by $g_t(0) = 0$ and $g_t'(0) > 0$. Moreover,

$$g_t'(0) = \exp\left(\int_0^t \mu_t([0, 2\pi))\mathrm{d}t\right). \tag{5.25}$$

In particular, if μ_t is a family of probability measures, then Ω_t form a Loewner chain with the canonical parametrization.

In the case when μ_t is a family of probability measures, the integral in the Loewner equation gives a function p_t from the Carathéodory class. So, alternatively, the above theorem could be reformulated in terms of p_t.

If $p_t(z)$ is continuous with respect to t and $p_t(z) \in \mathcal{P}$ for every t, then the conclusions of the theorem hold if g_t is a solution of the differential equation

$$\dot{g}_t(z) = g_t(z) p_t(g_t(z)), \quad g_0(z) = z. \tag{5.26}$$

This form also explains why we consider the existence of solutions only inside $\mathbb{D}$. When $|g_t(z)| \geq 1$, the right-hand side is not even defined and so solutions obviously cease to exist. In the integral form, this is less obvious, so we added it as an explicit condition.

In the case when $\mu_t([0, 2\pi)) \neq 1$, almost nothing changes. We can still consider a family of

$$\tilde{p}_t(z) = \int_0^{2\pi} \frac{e^{i\theta} + z}{e^{i\theta} - z} \mu_t(\mathrm{d}\theta).$$

These functions are not from the Carathéodory class, but they can be written as $\tilde{p}_t(z) = \mu_t([0, 2\pi)) p_t(z)$, where $p_t \in \mathcal{P}$. In this case, equation (5.25) holds with p_t replaced by $\tilde{p}_t$. This leads to a Loewner chain with non-canonical time parametrization. This could be remedied by an appropriate time change.

We state and prove the theorem under the assumption that the family μ_t (or p_t) is continuous with respect to t. In fact, it is sufficient to assume that the family is only measurable. The proof stays essentially the same, but one needs more general results of differential equations with measurable right-hand sides.

There are two slightly different methods of proving this theorem. One possibility is to rewrite the Loewner evolution as an integral equation:

$$g_t(z) = g_0(z) \exp\left(\int_0^t \tilde{p}_s(g_s(z)) \mathrm{d}s \right) = z \exp\left(\int_0^t \tilde{p}_s(g_s(z)) \mathrm{d}s \right).$$

This integral equation can be solved using the Picard–Lindelöf iteration method. Starting with $g_t^{(0)}(z) = 0$, we define

$$g_t^{(n+1)}(z) = z \exp\left(\int_0^t \tilde{p}_s(g_s^{(n)}(z)) \mathrm{d}s \right)$$

and show that $g_t^{(n)}$ converges to the solution of the Loewner evolution and derive its properties in the process.

An alternative approach, which we follow here, is to use the theory of differential equations to show that the local solutions exist and use the integral form to study their properties.

Proof. The main idea of the proof is rather straightforward. Since $p_t(z)$ is analytic in z, the solution $g_t(z)$ depends analytically on the initial condition

$g_0(z) = z$. Since trajectories started at two different points cannot merge before hitting a singularity, $g_t(z) \neq g_t(w)$ if $z \neq w$ and both solutions exist up to some time which is greater than t. This proves that g_t is univalent. To show that it is onto $\mathbb{D}$, we note that for any $w \in \mathbb{D}$ the 'reversed time trajectories' exist forever. Running them backwards from w, we can find a starting point z such that $g_t(z) = w$. In the following, we give the technical details to support these main ideas.

The local existence of solutions follows from standard results of ordinary differential equations. Local existence implies that $T(z) > 0$ and that our theorem is not vacuously true.

Equation (5.26) gives us

$$\partial_t \log |g_t(z)| = \operatorname{Re} \tilde{p}_t(g_t(z)) > 0.$$

This implies that $|g_t(z)|$ is increasing and we can see that

$$T(z) = \sup\{t : |g_t(z)| < 1\}.$$

Let us denote by $w = w(t)$ the trajectory $g_t(z_1)$ and $v = g_t(z_2)$. Assume that both solutions exist at least up to time t. By the Loewner equation, we have

$$\begin{aligned}\partial_t(w - v) &= \int_0^{2\pi} w \frac{e^{i\theta} + w}{e^{i\theta} - w} \mu_t(\mathrm{d}\theta) - \int_0^{2\pi} v \frac{e^{i\theta} + v}{e^{i\theta} - v} \mu_t(\mathrm{d}\theta) \\ &= (w - v) \int_0^{2\pi} \frac{e^{2i\theta} + e^{i\theta}(w + v) - wv}{(e^{i\theta} - w)(e^{i\theta} - v)} \mu_t(\mathrm{d}\theta)\end{aligned}$$

Integrating this equation, we get

$$w - v = (z_1 - z_2) \exp\left(\int_0^t \left(\int_0^{2\pi} \frac{e^{2i\theta} + e^{i\theta}(w + v) - wv}{(e^{i\theta} - w)(e^{i\theta} - v)} \mu_s(\mathrm{d}\theta) \right) \mathrm{d}s \right).$$

Assuming that both w and v are away from the unit circle, say, both have modulus at most $1 - \varepsilon$ for all $s \leq t$, and we see that the integrand is uniformly bounded and $w - v = 0$ if and only if $z_1 = z_2$. This proves that $g_t(z)$ is one-to-one as a function of z. The same uniform bound allows one to pass to the limit as $z_2 \to z_1$ and show that $g_t(z)$ is differentiable with respect to z and

$$g_t'(z) = \exp\left(\int_0^t \left(\int_0^{2\pi} \frac{e^{2i\theta} + e^{i\theta} 2g_t(z) - g_t^2(z)}{(e^{i\theta} - g_t(z))^2} \mu_s(\mathrm{d}\theta) \right) \mathrm{d}s \right).$$

The last formula might be not very illuminating, but it is much simpler in terms of functions p_t. We are not giving the complete derivation, but assuming that $g_t'(z)$ exists, we can differentiate (5.26) to obtain

$$\dot{g}_t'(z) = g_t'(z)\left[\tilde{p}_t(g_t(z)) + g_t(z)\tilde{p}_t'(g_t(z))\right]$$

which after integration becomes

$$g_t'(z) = \exp\left(\int_0^t \tilde{p}_s(g_s(z)) + g_s(z)\tilde{p}_s'(g_s(z))\mathrm{d}s\right).$$

This is essentially the same formula for $g_t'(z)$ but written in terms of p_t instead of μ_t.

To analyze the behaviour at $z = 0$, we note that $g_t(0) = 0$ is a solution of the Loewner differential equation. Since we already know that the solutions of the initial value problems are unique, we get $g_t(0) = 0$.

Plugging this into the formula for $g_t'(z)$, we obtain

$$\begin{aligned} g_t'(0) &= \exp\int_0^t \tilde{p}(0)\mathrm{d}s = \exp\left(\int_0^t\int_0^{2\pi} \mu_s(\mathrm{d}\theta)\mathrm{d}s\right) \\ &= \exp\left(\int_0^t \mu_s([0,2\pi))\mathrm{d}s\right) \end{aligned}$$

which is exactly (5.25).

Finally, we want to show that $g_t(z)$ is onto $\mathbb{D}$. For fixed $w \in \mathbb{D}$ and $t > 0$, we want to find z such that the trajectory started at z will end up at w at time t. To find the starting point z, we can look at the 'backwards' trajectory started from w, namely, we consider the solution of the following initial value problem:

$$\dot{h}_s(w) = -h_s(w)\int_0^{2\pi} \frac{e^{i\theta} + h_s(w)}{e^{i\theta} - h_s(w)}\mu_{t-s}(\mathrm{d}\theta), \quad h_0(w) = w.$$

This is almost the same equation as the ordinary Loewner equation, but the time runs 'backwards' from t to 0. By the same argument as before, $|h_s(w)|$ is decreasing and the solution exists for all $s \in [0,t]$. Let us take $z = h_t(w)$ and consider $\tilde{g}_s(z) = h_{t-s}(w)$. By definition, $\tilde{g}_0(z) = h_t(w) = z$ and $\tilde{g}_t(z) = h_0(w) = w$. It is also immediately clear and can be verified by differentiation that this function $\tilde{g}_s$ is a solution to the Loewner differential equation. This proves that $g_t(h_t(w)) = w$ and $g_t(\Omega_t) = \mathbb{D}$. □

Since g_t satisfies the Loewner differential equation, we can easily show that the inverse function $f_t = g_t^{-1}$ satisfies

$$\dot{f}_t(z) = -zf_t'(z)\int_{\partial\mathbb{D}} \frac{\zeta + z}{\zeta - z}\mu_t(\mathrm{d}\zeta) = -zf_t'(z)\tilde{p}_t(z).$$

Another useful observation is that if f_t and g_t are as in Theorem 5.17, then the transition function $\phi_{s,t}(z) = g_s(f_t(z))$ also satisfies the Loewner differential equations. By explicit differentiation, we see that

$$\begin{aligned}\partial_t\phi_{s,t}(z) &= -z\phi_{s,t}'(z)\tilde{p}_t(z) = -z\phi_{s,t}'\int \frac{\zeta + z}{\zeta - z}\mu_t(\mathrm{d}\zeta),\\ \partial_s\phi_{s,t}(z) &= \phi_{s,t}(z)\tilde{p}_s(\phi_{s,t}(z)) = \phi_{s,t}(z)\int \frac{\zeta - \phi_{s,t}(z)}{\zeta - \phi_{s,t}(z)}\mu_s(\mathrm{d}\zeta).\end{aligned} \tag{5.27}$$

Exercise 5.9. Let $\mu_t(\mathrm{d}\theta) = m_t\mathrm{d}\theta/2\pi$, where m_t is a continuous bounded function of t. Solve the Loewner evolution driven by this measure.

Exactly as in the slit case, it is possible to formulate the external and the whole-plane versions of this theorem. The proof of the external version or its derivation from the interior version is absolutely straightforward. The whole-plane version is a bit more involved: the initial condition becomes $e^t g_t(z) \to z$ as $t \to -\infty$ and one has to show that it makes sense. The details can be found in Lawler (2005).

As before, the chordal case is very similar to the radial case with only minor technical differences.

Theorem 5.18. *Let μ_t, $t \in [0,\infty)$, be a family of non-negative Borel measures on $\mathbb{R}$ which is locally uniformly bounded in the sense that for every $T > 0$, there is $R = R(T) > 0$ such that the compact support of μ_t is inside $[-R,R]$ for all $t \leq T$ and*

$$\sup_{0\leq t\leq T} \mu_t(\mathbb{R}) < \infty.$$

We also assume that μ_t is continuous with respect to t in the weak topology. For each $z \in \mathbb{H}$, let $g_t(z)$ be the solution of the chordal Loewner differential equation

$$\dot{g}_t(z) = \int_{\mathbb{R}} \frac{\mu_t(\mathrm{d}x)}{g_t(z) - x}$$

with the initial condition $g_0(z) = z$. By $T(z)$, we denote the supremum of t such that the $g_s(z)$ exists and is inside $\mathbb{H}$ for all $s < t$. By K_t, we denote *the set of points z such that $T(z) \le t$ and by $\mathbb{H}_t = \mathbb{H} \setminus K_t$ the set of points where solution exists up to time t.*

Then $g_t(z)$ is the unique conformal map from $\mathbb{H}_t$ onto $\mathbb{H}$ with thermodynamic normalization at infinity

$$g_t(z) = z + b(t)z^{-1} + \cdots .$$

Moreover,

$$b(t) = \int_0^t \mu_s(\mathbb{R})\mathrm{d}s.$$

Proof. The proof follows the same plan as in the radial case. First, we note that

$$\partial_t(\operatorname{Im} g_t(z)) = -\int_{\mathbb{R}} \frac{\operatorname{Im} g_t(z)}{|g_t(z) - x|^2}\, \mu_t(\mathrm{d}x) < 0.$$

Hence the imaginary part of $g_t(z)$ is decreasing with time and $T(z) = \sup\{t : \operatorname{Im} g_t(z) > 0\}$. In other words, exactly as in the radial case, points move closer to the boundary.

As before, we consider w and v to be trajectories of two distinct points z_1 and z_2 and assume that both trajectories exist up to time t. By the chordal Loewner differential equation,

$$\partial_t(w - v) = -(w - v)\int_{\mathbb{R}} \frac{\mu_t(\mathrm{d}x)}{(w - x)(v - x)}.$$

This formula after integration gives

$$w - v = (z_1 - z_2)\exp\left(-\int_0^t \left(\int_{\mathbb{R}} \frac{\mu_s(\mathrm{d}x)}{(w - x)(v - x)}\right) \mathrm{d}s\right).$$

This immediately gives that while trajectories are bounded away from the real line, they do not merge. Moreover, dividing by $(z_1 - z_2)$ and passing to the limit, we get

$$g_t'(z) = \exp\left(-\int_0^t \left(\int_{\mathbb{R}} \frac{\mu_s(\mathrm{d}x)}{(g_s(z) - x)^2}\right) \mathrm{d}s\right).$$

In particular, g_t is differentiable with respect to z.

The proof that g_t is onto $\mathbb{H}$ uses the same trick. Let us fix $w \in \mathbb{H}$ and $t \ge 0$. We run the backwards evolution

$$\dot{h}_s(w) = -\int_{\mathbb{R}} \frac{\mu_{t-s}(\mathrm{d}x)}{h_t(w) - x}, \quad h_0(w) = w.$$

Since the imaginary part of h_s is increasing, the solution exists up to time t. Taking $z = h_t(w)$, we can show in exactly the same way as before that $g_t(z) = g_t(h_t(w)) = w$.

Finally, we have to show that g_t has the right normalization at infinity. For large z, using that the measures μ_t are locally uniformly bounded, we have

$$\dot{g}_t(z) = \int_{\mathbb{R}} \frac{\mu_t(\mathrm{d}x)}{g_t(z) - x} = \frac{\mu_t(\mathbb{R})}{g_t(z)} + o\left(\frac{1}{g_t(z)}\right) = \frac{\mu_t(\mathbb{R})}{z} + o\left(\frac{1}{z}\right).$$

Integrating, we obtain

$$g_t(z) = z + \left(\int_0^t \mu_s(\mathbb{R})\mathrm{d}s\right)\frac{1}{z} + \cdots .$$

which is exactly the thermodynamic normalization. □

Exercise 5.10. Let μ_t be a family of measures as in the above theorem. Let us consider the measures shifted by $x_0 \in \mathbb{R}$, that is, $\tilde{\mu}_t(\mathrm{d}x) = \mu_t(\mathrm{d}x - x_0)$. Show that the corresponding hulls $\tilde{K}_t$ are given by $\tilde{K}_t = K_t + x_0$. This statement is the measure-driven counterpart of Exercise 5.6.

Exercise 5.11. Let K_t be generated by the chordal Loewner evolution driven by measures μ_t. Assume that K_t are parametrized by capacity; equivalently $\mu_t(\mathbb{R}) = 2$ for all t. Show that $\operatorname{Im} z \le 2\sqrt{t}$ for every $z \in K_t$.

Exercise 5.12. Let K be a half-plane hull containing point iy_0. Show that $\operatorname{hcap} K \ge y_0^2/2$. This means that out of all hulls containing a given point, the vertical interval has the smallest capacity.

Exercise 5.13. Let μ_t be a family of measures such that $\operatorname{supp} \mu_t \subset [a, b]$ and $\mu_t(\mathbb{R}) = 2$ for all t. Let K_t be half-plane hulls generated by the chordal Loewner evolution driven by μ_t. Show that $K_t \subset [a, b] \times [0, 2\sqrt{t}]$.

It is clear that the Loewner equation for g_t is much easier to analyze, as it is an ordinary differential equation, unlike a partial differential equation for f_t. Unfortunately, in many problems, we need to understand f_t and extracting information about f_t from the information about g_t can be difficult or even virtually impossible. The trick that we used in the proofs of Theorems 5.17 and 5.18 can be used to bypass this problem to some extent.

Given a family of measures μ_s, $0 \le s \le t$, on either the unit circle or on the real line, we define the *backward Loewner Evolutions* h_s as

$$\dot{h}_s(w) = -h_s(w) \int_0^{2\pi} \frac{e^{i\theta} + h_s(w)}{e^{i\theta} - h_s(w)} \mu_{t-s}(\mathrm{d}\theta), \quad h_0(w) = w \tag{5.28}$$

and

$$\dot{h}_s(w) = -\int_{\mathbb{R}} \frac{\mu_{t-s}(\mathrm{d}x)}{h_t(w) - x}, \quad h_0(w) = w. \tag{5.29}$$

As we have explained above, $h_t = g_t^{-1} = f_t$ where g_t and f_t are the corresponding functions defined by 'forward' Loewner evolutions driven by μ_s.

The family of maps, h_s, defines a family of shrinking domains $h_s(\mathbb{D}) = \tilde{\mathbb{D}}_s \subset \mathbb{D}$ in the radial case and $h_s(\mathbb{H}) = \tilde{\mathbb{H}}_s \subset \mathbb{H}$ in the chordal case. In both cases, the final domains $\tilde{\mathbb{D}}_t$ and $\tilde{\mathbb{H}}_t$ are the same as $f_t(\mathbb{D})$ and $f_t(\mathbb{H})$ respectively. But this is not true for intermediate values $0 < s < t$. In particular, it could be that the image of f_s is a slit domain for $s < t$, but the image of h_s is not. The first example of this type was constructed by Kufarev (1947).

Let $\kappa = \kappa(t) = e^{-t} + i\sqrt{1 - e^{-2t}}$ and $\lambda(t) = \kappa^3(t)$. Then the solution to the radial backward Loewner evolution driven by $\lambda(t)$ is

$$\frac{\kappa}{\kappa^2 + 1}(z + \kappa^2 - \kappa^2\sqrt{(1-z)(1-z/\kappa^4)}),$$

which is the normalized map from the unit disc onto the domain which is the connected component containing the origin of the unit disc without a circular arc passing through κ and λ and orthogonal to the unit circle.

Here, we present a half-plane version of this example introduced in Kager *et al.* (2004). Let $\gamma(t)$ be a semicircle of radius 1 centred at the origin. We assume that it grows from 1 to -1. We also assume that $\gamma(t)$ is parametrized by capacity. Since the half-plane capacity of the unit half-disc is 1, we immediately see that $t \in [0, 1/2)$. The domain $\mathbb{H}_t = \mathbb{H} \setminus \gamma([0, t])$ is the upper half-plane with a slit along a circular arc which is orthogonal to the real line (see Figure 5.5). This makes it a particularly simple domain, so the corresponding mapping-out function g_t can be explicitly computed in terms of $\gamma(t)$. Exercise 3.18 gives the map g_t in terms of θ, where $e^{i\theta} = \gamma(t)$, and also gives the capacity in terms of θ. Matching two parametrizations, we have $1 - \cos^4(\theta/2) = 2t$, which gives $\cos^2(\theta/2) = \sqrt{1-2t}$. Plugging this into the formula for g_θ, we get

$$g_t(z) = \frac{2(z-1) + 4z\sqrt{1-2t} + 2(z+1)\sqrt{(z+1)^2 - 4z\sqrt{1-2t}}}{4z}.$$

The same computation gives that the image of the tip $\gamma(t)$ is

$$u(t) = 3\sqrt{1-2t} - 2.$$

By Theorem 5.8, or by direct differentiation, g_t is a solution of

$$\dot{g}_t(z) = \frac{2}{g_t(z) - u(t)}, \quad g_0(z) = z.$$

It is important to understand what happens when t approaches 1/2. By passing to the limit, it is easy to see that

$$\lim_{t\to 1/2} g_t(z) =: g_{1/2}(z) = z + \frac{1}{z},$$

which is the Joukowsky map. This should not be a surprise. Indeed, if we consider $f_t = g_t^{-1}$, then they map the upper half-plane onto the half-plane with a cut. As $t \to 1/2$, the cut becomes a half-circle. Domains $\mathbb{H}_t$ converge to $\mathbb{H}_{1/2} = \mathbb{H} \setminus \mathbb{D}$ in the Carathéodory sense. By the half-plane version of the Carathéodory convergence theorem, the corresponding maps converge as well.

Let us now fix $t = 1/2$ and try to understand what the corresponding backward map, h_s, looks like. It should be the solution of

$$\dot{h}_s(z) = -\frac{2}{h_s(z) - u(1/2 - s)} = -\frac{2}{h_s(z) - (3\sqrt{2s} - 2)}, \quad h_0(z) = z.$$

Let us consider $H_s(z) = h_s(z-2) + 2$. By differentiation with respect to s, we can see that it satisfies the slightly simpler Loewner differential equation

$$\dot{H}_s(z) = -\frac{2}{H_s(z) - 3\sqrt{2s}}.$$

This equation can be solved explicitly. The solution is

$$H_s(z) = \sqrt{2s} + \frac{z}{2} + \frac{1}{2}\sqrt{z^2 - 4\sqrt{2s}z}.$$

Some algebraic manipulation shows that indeed

$$\dot{H}_s(z) = \frac{1}{\sqrt{2s}} - \frac{z}{\sqrt{2s}\sqrt{z^2 - 4\sqrt{2s}z}}$$

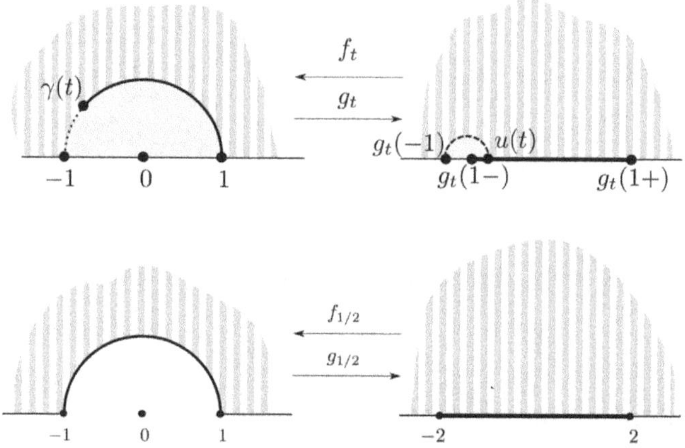

Fig. 5.5. Top: For $t < 1/2$, a circular arc is mapped out. Its tip is mapped to $u(t) = 3\sqrt{1-2t}-2$. The unit half disc (solid grey) is mapped to a small half disc; its endpoints are $u(t)$ and $g_t(-1) = \sqrt{1-2t}-2$. The dashed line, which is part of the unit circle and not mapped out yet, is mapped to the boundary of this small half disc.
Bottom: For $t = 1/2$, the domain is no longer a slit domain; it becomes the complement of the unit half disc.

is equal to

$$-\frac{2}{H_s(z) - 3\sqrt{2s}} = \frac{4}{4\sqrt{2s} - z - \sqrt{z^2 - 4\sqrt{2s}z}}.$$

Yet another computation shows that H_s is the thermodynamically normalized map from $\mathbb{H}$ onto itself without the half disc of radius $\sqrt{2s}$ centred at $2\sqrt{2s}$. The simplest way is to compute this map which is just a composition of two linear maps with the inverse of the Joukowsky function and to verify that it is indeed the expression in the definition of H_s. This implies that h_s maps onto the complement of the half-disc of the same radius $\sqrt{2s}$ centred at $2\sqrt{2s}-2$. In the limit $s = 1/2$, we get the half-disc of radius 1 centred at the origin. This means that $h_{1/2} = f_{1/2} = g_{1/2}^{-1}$, exactly as we should expect. For $t < 1/2$ maps, h_t and f_t are different (see Figures 5.5 and 5.6).

5.4. When a Loewner Chain is Generated by a Curve?

We have seen that in the case when the family of domains is generated by cutting out a Jordan curve from the base domain (the unit disc in the radial case and the upper half-plane in the chordal case), then the Loewner evolution equation is particularly simple. In this case, the driving measures

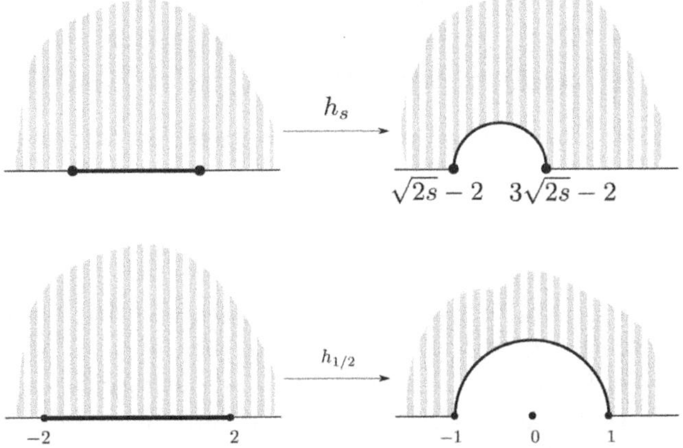

Fig. 5.6. Top: For $t = 1/2$ and for $s < t$, the backward flow function h_s maps the upper half-plane onto the complement of a half disc of radius $\sqrt{2s}$ which intersects the real line at $\sqrt{2s} - 2$ and $2\sqrt{2s} - 2$. Note that after the time change $\tau = 1/2 - s$, this is the same half disc which is the image of the unit half disc under the map g_τ (see Figure 5.5). Bottom: For $s = 1/2$, the backward flow map becomes $h_{1/2} = f_{1/2} = g_{1/2}^{-1}$.

are δ-measures and we can talk about the driving function which gives the position of the δ-measure.

But at the end of the last section, we saw an example when the Loewner evolution is given by a driving function but the corresponding curve hits the boundary and at this moment the corresponding domain stops being a slit domain. This means that these two properties are not really equivalent. In fact, there are three seemingly related properties: Loewner evolution is driven by a continuous function; the hulls are generated by a continuous curve; and the Loewner evolution describes a slit domain. It is an interesting and important problem to understand how these properties are related and how to reformulate them in other terms.

Loewner chains driven by continuous functions can be characterized by a simple geometric condition. This was done by Pommerenke (1966). Following his work, we state the theorem in the radial case but essentially the same result with essentially the same proof holds in the chordal case.

Theorem 5.19. *Let $f_t(z)$ be a bounded normalized Loewner chain defined for $0 \le t \le T \le \infty$ and Ω_t be the corresponding domains. Then the following two conditions are equivalent.*

(1) *There is a continuous function $\lambda(t)$ with $|\lambda(t)| = 1$ such that f_t and $g_t = f_t^{-1}$ satisfy the Loewner differential equations driven by $\lambda(t)$.*

(2) *For every* $\varepsilon > 0$, *there is* $\delta > 0$ *such that for all* $0 \le s < t \le T$ *with* $(t - s) \le \delta$, *there is a cross-cut of* Ω_s *with diameter at most* ε *which separates* $\Omega_s \setminus \Omega_t$ *from* 0 *inside* Ω_s.

Remark 5.20. The assumption of boundedness is not essential. If we allow domains to be unbounded, then in the second condition we should use the spherical metric instead of the Euclidean.

Remark 5.21. On a hand-waving level, this theorem is very natural and rather obvious. Indeed, by Theorem 5.16, f_t and g_t satisfy measure-driven Loewner differential equations. Corresponding measures, μ_t, tell us how fast the boundary of Ω_t moves. The fact that μ_t is a delta-measure at λ_t gives us essentially that the boundary changes at only one place. The second condition is almost the same as the statement that $\bigcap_{t>s}(\Omega_s \setminus \Omega_t)$ is a single prime-end, which also means that, infinitesimally, domains change at one place only. Unfortunately, making these hand-waving arguments rigorous is not so simple.

Proof of Theorem 5.19. We start with the simpler implication (2) $\Rightarrow$ (1). The proof is rather simple. It is essentially based on the observation that much of the proof of Theorem 5.4 is independent from the fact that Ω_t are slit domains.

To be consistent with the notation of Section 5.1.1, we define

$$\gamma_{s,t} = g_s(\Omega_s \setminus \Omega_t) = \mathbb{D} \setminus \phi_{s,t}(\mathbb{D}).$$

Note that this notation might be a bit misleading: we no longer assume that $\gamma_{s,t}$ is a Jordan arc. It is clear from the definition that $\gamma_{s,t} \subset \gamma_{s,t'}$ for $t' > t$.

Let $s_n \le t_n$ and $t_n - s_n \to 0$. By assumption (2), there are cross-cuts c_n separating $\Omega_{s_n} \setminus \Omega_{t_n}$ from 0 such that $\operatorname{diam}(c_n) \to 0$. These cross-cuts are uniformly bounded away from 0. Let $\omega(0, c_n) = \omega(c_n)$ be the harmonic measure of c_n in the domain which is Ω_{s_n} without the interior of c_n (that is we cut Ω_{s_n} using the cross-cut c_n and take the component containing 0). By monotonicity of the harmonic measure and the Beurling estimate (Corollary 4.17), $\omega(c_n) \le c \operatorname{diam}(c_n)^{1/2}$, where c is a constant. In particular, this implies that the harmonic measure of $f_{s_n}(c_n)$ in $\mathbb{D} \setminus f_{s_n}(c_n)$ uniformly tends to zero. As in the proof of Theorem 4.21, Proposition 4.19 implies that in this case $\operatorname{diam}(f_{s_n}(c_n)) \to 0$. Since f_{s_n} are cross-cuts in $\mathbb{D}$ separating γ_{s_n,t_n} from the origin, this implies that $\operatorname{diam}(\gamma_{s_n,t_n}) \to 0$. This allows us to define $\lambda(s)$ to be the limit of $\gamma_{s,t}$ as $t \to s$.

Note that in the proof of Lemma 5.2, we use that γ is a curve only in one place: when we argued that $\gamma_{s,t}$ shrinks to $\lambda(s)$. Since we already know that this is true in our case as well, the rest of the proof is exactly the same and we have that $\lambda(s)$ is continuous. Since the proof of the Loewner theorem for Jordan curves used only the fact that λ is continuous, the same argument is valid in our case and we obtain that f_t and g_t solve the Loewner differential equations driven by a continuous function λ.

To prove the other direction (1) $\Rightarrow$ (2), we will show that for t sufficiently close to s and z away from $\lambda(t)$, the image $\phi_{s,t}(z)$ stays away from $\lambda(s)$. Using the Loewner evolution, this will imply that $\phi_{s,t}(z)$ is close to z and the diameter of $\gamma_{s,t}$ is small. By the length–area lemma, this will allow us to construct a short cross-cut as in condition (2).

To make this argument rigorous, we start by considering some small $r > 0$ and choose $0 < \delta < r^2/32$ sufficiently small so that

$$|\lambda(s) - \lambda(t)| \leq r/4$$

for all $s \leq t$ with $t - s \leq \delta$. Let $z \in \mathbb{D}$ such that $|\lambda(t) - z| \geq r$. We claim that $|\lambda(s) - \phi_{s,t}(z)| > r/2$ for all $s \in [t - \delta, t]$.

Suppose that this is not true. Then by continuity, there is a largest $s' \in [t - \delta, t]$ such that $|\lambda(s') - \phi_{s',t}(z)| = r/2$. Since $\phi_{s,t}$ is a solution of the Loewner differential equation, we have

$$\partial_s \phi_{s,t}(z) = \phi_{s,t} \frac{\lambda(s) + \phi_{s,t}(z)}{\lambda(s) - \phi_{s,t}(z)}.$$

By construction, $|\lambda(s) - \phi_{s,t}(z)| > r/2$ for all $s \in [s', t]$, which implies

$$|\phi_{s',t}(z) - z| = |\phi_{s',t}(z) - \phi_{t,t}(z)| < (t - s')\frac{2}{r/2} \leq \delta\frac{4}{r} \leq \frac{r}{8}. \tag{5.30}$$

Together with $|\lambda(t) - z| \geq r$, this gives us

$$|\lambda(t) - \phi_{s',t}(z)| > 7r/8.$$

But $|\lambda(s') - \phi_{s',t}(z)| = r/2$ and $|\lambda(s') - \lambda(t)| \leq r/4$ which imply

$$|\lambda(t) - \phi_{s',t}(z)| < 3r/4$$

and contradict the previous estimate. This proves that if $|\lambda(t) - z| > r$ then

$$|\lambda(s) - \phi_{s,t}(z)| > r/2 \qquad \text{for all } s \in [t - \delta, t]. \tag{5.31}$$

Since this lower bound holds for all $s \in [t - \delta, t]$, the estimate (5.30) also holds for all $s \in [t - \delta, t]$.

Let us consider a circular arc $\{z \in \mathbb{D} : |z - \lambda(t)| = r\}$. The image of this arc under $\phi_{s,t}$ is close to $\lambda(s)$ for all $s \in [t-\delta, t]$. Indeed, for any z from this arc

$$|\lambda(s) - \phi_{s,t}(z)| \le |\lambda(s) - \lambda(t)| + |\lambda(t) - z| + |z - \phi_{s,t}(z)| < \frac{3r}{2}.$$

Finally, we claim that the image of this arc is a cross-cut, that is, its end-points are mapped to the unit circle.

We want to show that $\phi_{s,t}(z)$ is close to the unit circle when $|z|$ is close to 1, namely, we claim that $|\phi_{s,t}(z)| > |z|^2$ for all $s \le t \le s + \delta$. Again, we prove it by contradiction. Let $s < t' < s + \delta$ be the first time such that $|\phi_{s,t'}(z)| = |z|^2$. Since ϕ satisfies the Loewner differential equations (5.27),

$$\partial_t \log|\phi_{s,t}(z)| = \operatorname{Re}\left(\frac{\lambda(t) + \phi_{s,t}(z)}{\lambda(t) - \phi_{s,t}(z)}\right) = \frac{1 - |\phi_{s,t}(z)|^2}{|\lambda(t) - \phi_{s,t}(z)|^2}.$$

Using this identity and the inequality (5.31), we get

$$\begin{aligned} \log(1/|z|) &= \left|\log|z|^2 - \log|z|\right| = \left|\log|\phi_{s',t}(z)| - \log|\phi_{t,t}(z)|\right| \\ &\le (t - s')\frac{1 - |z|^4}{(r/2)^2} \le \frac{r^2}{32}\frac{4}{r^2}(1 - |z|^4) \le \frac{1}{2}\frac{1 - |z|^4}{4} \\ &\le \frac{1}{2}\log(1/|z|). \end{aligned}$$

This contradiction implies that $|\phi_{s,t}(z)| > |z|^2$. If we pass to the limit as $|z| \to 1$, we see that $|\phi_{s,t}(z)| \to 1$. This implies that the image of the above-considered circular arc is indeed a cross-cut. Moreover, the diameter of this cross-cut is at most $3r/2$. Let us denote this cross-cut by c_r.

We denote $K_{s,t} = \Omega_s \setminus \Omega_t$ and note that $K_{s,t} = f_s(\gamma_{s,t})$. The above argument implies that for every $r > 0$, there is $\delta > 0$ such that if $t - s < \delta$ then $\gamma_{s,t}$ is in $3r/2$ neighbourhood of $\lambda(s)$. Let us consider circular arcs $c_\rho = \{z \in \mathbb{D} : |z - \lambda(s)| = \rho\}$ with $3r/2 < \rho < 1$. All these arcs separate 0 from $\gamma_{s,t}$ and their images separate $K_{s,t}$ from the origin. Applying the length–area lemma (Lemma 2.22), we get that there is $\rho \in [3r/2, 1]$ such that

$$\operatorname{length}(f(c_\rho)) \le \frac{\sqrt{2\pi \operatorname{Area}(\Omega_s)}}{\log(2/3r)}.$$

For every ε, we can choose r sufficiently small, so that

$$\frac{\sqrt{2\pi \operatorname{Area}(\Omega_s)}}{\log(2/3r)} < \varepsilon.$$

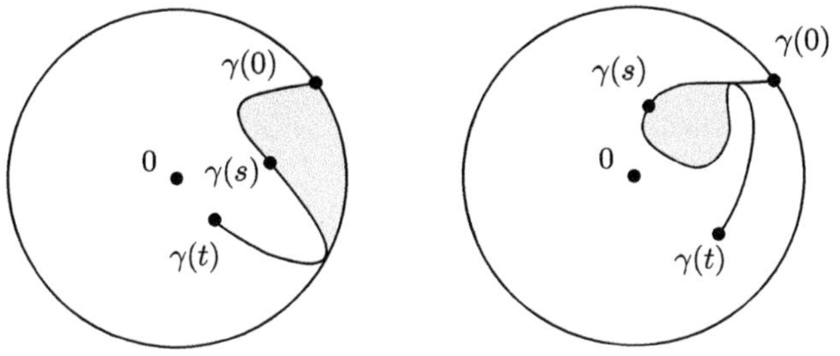

Fig. 5.7. Examples when K_t is not equal to $\gamma([0,t])$. On the left, the curve is touching the boundary; on the right, it has a double point but it does not cross itself.

By choosing the corresponding δ, we obtain that for every $t - s < \delta$, there is a cross-cut separating $K_{s,t}$ from 0 which has length (and diameter) at most ε. This is exactly condition (2). □

This theorem and the example at the end of the previous section show that there are domains that are not given by cutting away a Jordan arc, which are nevertheless described by the Loewner evolution driven by a continuous function.

Before moving forward, let as consider some examples that are not slit domains but satisfy condition (2) of Theorem 5.19.

Let $\gamma(t) : [0,T] \to \bar{\mathbb{D}}$ such that $\gamma(0) = 1$ and $\gamma(t) \neq 0$. Define Ω_t to be the component of $\mathbb{D} \setminus \gamma([0,t])$ which contains the origin. In this case, $K_t = \mathbb{D} \setminus \Omega_t$ is not necessary the same as $\gamma([0,t])$. If the curve is not simple or touches the unit circle, K_t is the 'hole-filled' curved, namely $\gamma([0,t])$ together with all components of its complement not containing the origin. In this case, we say that K_t and Ω_t are *generated by the curve* γ. The curve γ is also called the *trace* of the corresponding Loewner evolution. We formulate this in the radial case; in the chordal case, it is almost the same but we have to take the unbounded component of $\mathbb{H} \setminus \gamma([0,t])$.

Figure 5.7 shows two examples of such curves. In both cases, $K_s = \gamma([0,s])$ but K_t is the curve together with a grey bounded region. Domains Ω_τ satisfy condition (2) of Theorem 5.19; hence they are given by the Loewner evolution with a continuous driving function.

The left part of Figure 5.8 gives an example of a curve which crosses itself. In this case, the domains form a continuous (in the Carathéodory sense) family and can be parametrized by capacity and described by the Loewner evolution. Let $t_1 < t_2$ be times corresponding to the double point. Then for every δ, the difference between $\Omega_{t_2+\delta}$ and $\Omega_{t_2-\delta}$ is made of two

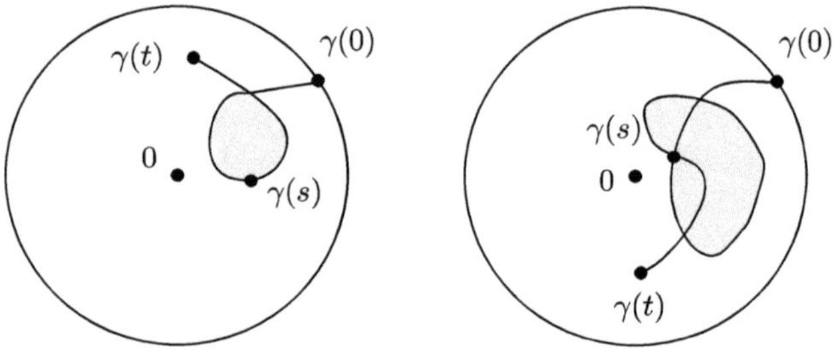

Fig. 5.8. On the left is the example of a 'hole-filled' curve which does not satisfy condition (2). On the right, K_t is not strictly increasing; it does not change while γ is inside a previously formed loop.

parts that are separated by $\gamma([0, t_2])$. One part is a short curve $\gamma([t_2 - \delta, t_2))$ together with the grey area, the other part is $\gamma((t_2, t_2 + \delta])$ which is on the other side of γ. The union of these two sets cannot be separated from 0 by a short cross-cut. Thus, condition (2) fails. However, it holds for all $t \neq t_2$. In this case, we still have an evolution driven by a function, but it has a jump discontinuity at time t_2. At time t_2, the image of the tip of the curve jumps from the left image of $\gamma(t_1)$ to the right image of $\gamma(t_1)$ under the map f_{t_2}.

On the right side of Figure 5.8, we have an example of a curve which re-enters a previously created loop. As a result, the hole-filled domains do not increase while the curve is inside the grey domain. If we parametrize $\gamma(t)$ using the standard parametrization of $\Omega_t = \mathbb{D} \setminus K_t$, then the curve is not continuous. It jumps from the point where it enters the loop to the point where it exits.

Considering these examples, it is clear that if K_t is generated by a curve γ, then the corresponding Loewner evolution is driven by a continuous function if and only if γ does not cross itself (but it can have double points).

Unfortunately, the next example shows that a Loewner evolution driven by a continuous function might not come from a curve at all. This example makes sense in both radial and chordal cases, but it is easier to draw a picture in the half-plane case.

In this example, we consider the family of growing compacts K_t that are constructed in the following way. For $t \in [0, 1)$, the set $K_t = \gamma([0, t])$, where γ is a curve which starts on the real line and goes like a logarithmic spiral around a disc slowly approaching its boundary (solid line on Figure 5.9). The set K_1 is the hole-filled closure of $\gamma([0, 1))$. In other words, it

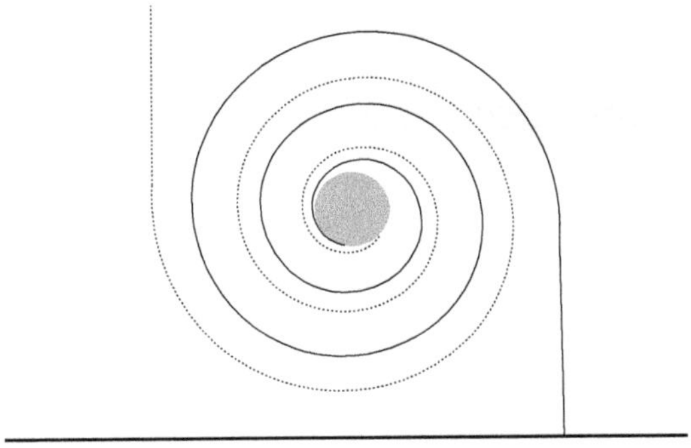

Fig. 5.9. Double spiral domain is not generated by a curve.

is the union of $\gamma([0,1))$ and the closed disc (grey disc on Figure 5.9). For $t > 1$, it is the union of K_1 and another spiral which is disjoint from K_1. We define Ω_t to be $\mathbb{H} \setminus K_t$. It is immediately obvious that condition (2) is satisfied when $1 \notin [s,t]$ because then $K_t \setminus K_s$ is a nice short Jordan arc. When the time is close to 1, we note that the distance between different revolutions of one spiral or between two spirals can be arbitrarily small, so there are arbitrarily short cuts separating the disc from infinity. This should not be too surprising since at time 1 the unit disc that we add might have a large diameter, but its boundary is the impression of a single prime-end (see Figure 2.6). This implies that the Loewner evolution describing Ω_t is driven by a continuous function. In fact, an even stronger result holds (Lind *et al.*, 2010; Marshall and Rohde, 2005): the driving function is Hölder-1/2 continuous. But since at time 1 we add the entire circle to the boundary, the sets K_t are not generated by a continuous curve. Another example of this type could be constructed by considering other growing compacts with boundaries which are not locally connected.

These examples suggest that the problem arises when the growth happens at a prime-end which is not an accessible point. When the boundary is always locally connected, the maps are continuous and it is natural to assume that hulls must be generated by a continuous curve. Surprisingly, this is not quite true. The positive part is given by the following theorem.

Theorem 5.22. *Let g_t be a chordal or radial Loewner evolution driven by a continuous function $u(t)$ and let K_t be the corresponding growing sets. Let us assume that K_t has a locally connected boundary for every t. Let us*

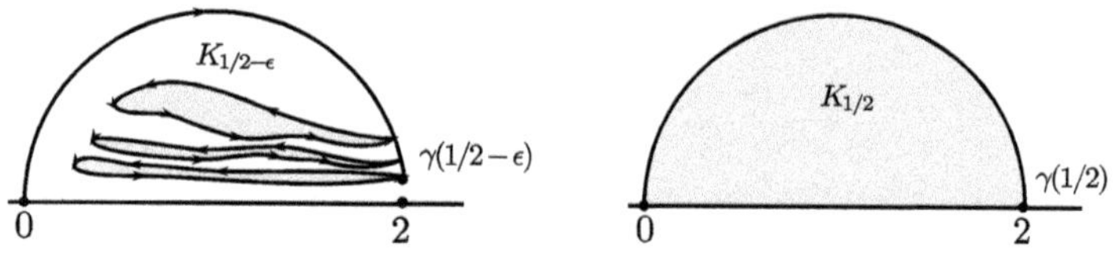

Fig. 5.10. An example of a curve which is not locally connected, only right continuous, but the corresponding Loewner chain is driven by a continuous driving function and the hulls are locally connected.

define

$$\gamma(t) = \lim_{y \to 0} g_t^{-1}(u(t) + iy). \tag{5.32}$$

Then $\gamma(t)$ is right continuous and the Loewner chain is generated by γ, namely, $\mathbb{H}_t = \mathbb{H} \setminus K_t$ is the unbounded component of the complement of the closure of $\gamma([0,t])$.

Before proving the theorem, we would like to mention that the curve γ does not have to be continuous. This can be seen from a simple counterexample shown in Figure 5.10. In this example, we start with a curve $\gamma(t)$ which we assume to be parametrized by capacity. For $t < 1/2$, this curve behaves like this: it starts as an arc of a semi-circle, then it makes a long excursion inside the unit disc, then it continues along the semi-circle, makes an excursion and so on. We assume that the points from which the curve makes an excursion accumulate towards 2; for simplicity, we assume that the nth excursion starts from a point which is 2^{-n} away from 2. We also assume that all excursions are of diameter at least $1/2$.

By comparing $\gamma([0,t])$ with the half-disc and a circular slit, we can see that $\text{hcap}(\gamma([0,t])) \nearrow 1$, and $\gamma([0,t])$ converges to the half-disc in the Carathéodory sense as $t \to 1/2$. This implies that it is indeed possible to parametrize this curve by capacity for $t \in [0, 1/2)$. As $t \to 1/2$, this curve is obviously not continuous. We define $\gamma(1/2) = 2$.

For all $s < t < 1/2$, the increment $\gamma([s,t])$ can be separated from infinity by a short cross-cut. This shows that this curve satisfies a half-plane version of condition (2) of Theorem 5.19. Although we prove it only in the disc case, it is not so difficult to show that it holds in the half-plane case as well. This implies that the corresponding Loewner chain is generated by a continuous function.

Finally, for all $t < 1/2$, K_t is generated by a very nice continuous curve; hence it is locally connected. At $t = 1/2$, the hull is the half-disc and also locally connected. This shows that the hulls K_t for $t \in [0, 1/2]$ satisfy all assumptions of Theorem 5.22, but the trace 'curve' γ is only right continuous, not left continuous. Moreover, the trace $\gamma([0,1/2])$ is itself not locally connected, but the hulls K_t are locally connected.

Proof of Theorem 5.22. We are going to state the proof in the chordal case. The proof in the radial case is almost exactly the same.

We start by noticing that since the boundary of $\mathbb{H}_t$ is locally connected, by Theorem 2.26 the function g_t^{-1} can be continuously extended to the boundary. This theorem is formulated for maps onto $\mathbb{D}$, but composing with an appropriate Möbius transformation, we get the same result for maps onto $\mathbb{H}$. This means that the limit in (5.32) is well defined.

From the proof of Theorem 5.19, we know that the diameter of $K_{s,t} = g_s(K_t \setminus K_s)$ for $s < t$ tends to zero as $t \to s$. Since g_s^{-1} is continuous, $\operatorname{diam}(K_t \setminus K_s) \to 0$ as well. This implies that $\gamma(t)$ is right continuous.

Since the boundary of K_t is locally connected and g_t^{-1} is continuous, the image of a vertical interval approaching $u(t)$ is a rectifiable curve defining an accessible point in $\mathbb{H}_t$. Hence, $\gamma(t)$ is an accessible point in $\mathbb{H}_t$. Moreover, this is the only accessible point in

$$X_t = K_t \setminus \bigcup_{s<t} K_s.$$

It remains to be shown that K_t is generated by $\bar{\gamma}$. To show this, it is sufficient to prove that the boundary of K_t is contained in $\overline{\gamma([0,t])}$. To be more precise, we need

$$\partial_t = ((\partial \mathbb{H}_t) \cap \mathbb{H}) \subset \overline{\gamma([0,t])}.$$

Let ζ be any point in ∂_t. By definition, for any $\varepsilon > 0$, there is $z \in \mathbb{H}_t$ such that $|z - \zeta| < \varepsilon$. Let us connect z and ζ by a straight interval and let ζ' be the first intersection of this interval with K_t. By construction, this is an accessible point on the boundary of K_t which is ε-close to ζ. Let $s = T_{\zeta'} \le t$. Since this point is accessible in $\mathbb{H}_t$, it is also accessible in $\mathbb{H}_s$. Since $\gamma(s)$ is the only accessible point in X_s, we have that $\zeta' = \gamma(s)$. This proves that ζ is in the closure of $\gamma([0,t])$. This completes the proof. □

This theorem gives us a condition in terms of either the geometry of K_t or analytic properties of f_t. It would be nice to have a necessary and sufficient condition in terms of the driving function. We also would like to know when the corresponding curve $\gamma(t)$ is simple. This question has been extensively studied but is still open. There are only sufficient conditions that are known to be not necessary and, vice versa, necessary but not sufficient conditions.

A useful sufficient condition is given by the following theorem by Lind (2005) (see also Marshall and Rohde (2005)).

Theorem 5.23. *Let g_t be the chordal Loewner evolution driven by $u(t)$ such that*

$$\sup_{s\neq t}\frac{|u(t)-u(s)|}{|t-s|^{1/2}}=\|u\|_{1/2}<4.$$

Then K_t is generated by a simple continuous curve γ.

The proof of this theorem is rather technical and we are not going to present it here. Instead, we would like to finish with a couple of concluding remarks.

It is known that the constant 4 in this theorem is sharp, that is, there is $u(t)$ with $\|u\|_{1/2}=4$ such that the corresponding curve is not simple or it could be that it is even not generated by a curve. But there are examples where K_t is a simple curve which is driven by $u(t)$ with arbitrarily large, or even infinite, norm $\|u\|_{1/2}$. Finally, we would like to mention that even if the curve γ does exist, it could be extremely complicated. In particular, it could even be a space-filling curve. These examples appear in the study of the Schramm–Loewner evolution, that is the Loewner evolution driven by a multiple of the Brownian motion. See Lawler (2005) for more information.

5.5. Solutions

Solution of 5.1. Since the curve is a straight slit, the map f_t must be a rescaling of the Joukowsky map. It is easy to see that this is indeed the case and that $f_t(z)=e^t(z+1/z+2)=e^t(1+z)^2/z$. The point 1 is always mapped to the tip of the slit; hence $\lambda(t)=1$. Plugging this into the right-hand side of the Loewner equation, we have

$$-zf_t'(z)\frac{1+z}{1-z}=-e^t\left(1-\frac{1}{z^2}\right)\frac{1+z}{1-z}=e^t\frac{(1+z)^2}{z}=\dot{f}_t(z).$$

The inverse function is

$$g_t(z)=e^{-t}\left(-e^t+\frac{1}{2}z+\frac{1}{2}\sqrt{z(z-4e^t)}\right).$$

It is easy to see that $e^tg_t(z)$, which is the last factor in the above formula, indeed converges to z as $t\to-\infty$. Explicit verification of the Loewner equation for g_t is left to the reader. □

Solution of 5.2. The Koebe function $K(z)=z/(1-z)^2=z+2z^2+\cdots$ maps the unit disc onto $\mathbb{C}\setminus(-\infty,-1/4]$; hence

$f_t(z) = -e^{-t}K(-z) = e^t z + \cdots$ maps the unit disc onto $\mathbb{C} \setminus [e^{-t}/4, +\infty)$. This gives the parametrization and the function f_t. The Koebe function maps -1 to the tip; hence f_t maps 1 to the tip. This implies that the driving function $\lambda(t)$ is identically equal to 1. Plugging this into the Loewner differential equation, we have

$$-zf_t'(z)\frac{1+z}{1-z} = -ze^{-t}\frac{1-z}{(1+z)^3}\frac{1+z}{1-z} = \dot{f}_t(z).$$

The inverse function is

$$g_t(z) = \frac{e^{-t} - 2z - e^{-t}\sqrt{1-4e^t z}}{2z}.$$

When $t \to -\infty$ and $|z|$ is uniformly bounded, it behaves as

$$g_t(z) = \frac{e^{-t} - 2z - e^{-t} + 2z + 2e^t z^2 + \cdots}{2z} = e^t z + O(e^{2t}).$$

This proves that $e^{-t}g_t(z) \to z$. □

Solution of 5.3. The function $cf_t(z)$ maps onto the complement of $c\gamma((-\infty, t])$ and has the expansion $ce^{-t}z + \cdots$ at the origin. This suggests considering the mapping

$$\tilde{f}_s(z) = cf_t(e^{-i\theta}z) = e^{-s}z + \cdots,$$

where $c = re^{i\theta}$ and $s = t - \log(r)$. The corresponding curve $\tilde{\gamma}(s) = c\gamma(t) = c\gamma(s + \log(r))$ is parametrized by capacity. This shows that the new time is just shifted by $\log(r)$. By explicit differentiation and the Loewner equation for f_t,

$$\partial_s(\tilde{f}_s(z)) = -cze^{-i\theta}f_t'(e^{-i\theta}z)\frac{\lambda(t) + ze^{-i\theta}}{\lambda(t) - ze^{-i\theta}} = -z\tilde{f}_s'(z)\frac{\lambda(t)e^{i\theta} + z}{\lambda(t)e^{i\theta} - z}.$$

This proves that the new driving function is $\tilde{\lambda}(s) = \lambda(t)e^{i\theta}$. □

Solution of 5.4. From Exercise 3.15, we know that the half-plane capacity of a vertical interval of length l is $l^2/2$. So, if we want our interval at time t to have capacity $2t$, its length should be $2\sqrt{t}$. This means that $\gamma(t) = 2i\sqrt{t}$.

The corresponding map has already been computed in Exercise 2.8: $g_t(z) = \sqrt{z^2 + 4t}$. Under this map, the tip $\gamma(t) = 2i\sqrt{t}$ is mapped to 0, so we have $u(t) = 0$. Differentiating g_t with respect to t, we find

$$\dot{g}_t(z) = \frac{1}{2}\frac{4}{\sqrt{z^2+4t}} = \frac{2}{\sqrt{z^2+4t}} = \frac{2}{g_t(z) - u(t)}.$$

□

Solution of 5.5. From Exercise 3.17, we know that the function

$$f(z) = (z + \alpha)^{1-\alpha}(z + \alpha - 1)^{\alpha}$$

maps $\mathbb{H}$ onto the half-plane with slit of length $\alpha^{\alpha}(1-\alpha)^{1-\alpha}$. Moreover, the capacity of this slit is $\alpha(1-\alpha)/2$, and $1-2\alpha$ is mapped to the tip of the slit. By the scaling property of capacity and mapping-out functions (see Proposition 3.26), we have that $f_t(z) = \lambda_t f(z/\lambda_t)$, where

$$\lambda_t = 2\sqrt{t}\alpha^{-1/2}(1-\alpha)^{-1/2}$$

maps $\mathbb{H}$ onto $\mathbb{H}$ with the slit of length

$$2\sqrt{t}\alpha^{\alpha-1/2}(1-\alpha)^{1/2-\alpha}.$$

Moreover, the capacity of this slit is $2t$ and the point $u(t) = (1-2\alpha)\lambda_t$ is mapped to the tip of the slit. This proves that

$$\gamma(t) = 2\sqrt{t}\alpha^{\alpha-1/2}(1-\alpha)^{1/2-\alpha}e^{i\alpha\pi}$$

is parametrized by capacity. Differentiating f_t with respect to t, we obtain

$$\begin{aligned}\dot{f}_t(z) &= f(z/\lambda_t)\partial_t\lambda_t + \lambda_t f'(z/\lambda_t)z\partial_t(1/\lambda_t)\\ &= (t\alpha(1-\alpha))^{-1/2} f'(z/\lambda_t)\frac{(z/\lambda_t+\alpha)(z/\lambda_t+\alpha-1)}{z/\lambda_t+2\alpha-1}\\ &\quad - \frac{z}{2t}f'(z/\lambda_t)\\ &= -f'(z/\lambda_t)\frac{2}{z-(1-2\alpha)\lambda_t} = -f_t'(z)\frac{2}{z-u(t)},\end{aligned}$$

where, in the second equality, we used explicit formulas for f, f', and λ_t. □

Solution of 5.6. It is easy to see that $\tilde{g}_t(z) = g_t(z-x)+x$ (see also Exercise 3.16). Differentiating with respect to t, we get

$$\partial_t\tilde{g}_t(z) = \frac{2}{g_t(z-x)-u(t)} = \frac{2}{\tilde{g}_t(z)-x-u(t)}.$$

This proves that the driving function of $\tilde{\gamma}_t$ is indeed $u(t)+x$. □

Solution of 5.7. Let us consider $\tilde{g}_t(z) = -\bar{g}_t(-\bar{z})$. The transformation $z \mapsto -\bar{z}$ preserves $\mathbb{H}$ and maps $\mathbb{H}\setminus\gamma([0,t])$ to $\mathbb{H}\setminus\tilde{\gamma}([0,t])$. It is a standard fact, which follows immediately from the Cauchy–Riemann equations, that this function is analytic. The normalization at infinity is easy to understand:

$$\tilde{g}_t(z) = -\overline{\left(-\bar{z} + \frac{2t}{-\bar{z}} + O(z^{-2})\right)} = z + \frac{2t}{z} + O(z^{-2}).$$

This implies that $\tilde{\gamma}$ is also parametrized by capacity and that $\tilde{g}_t$ is the corresponding mapping. Differentiating with respect to t, we have

$$\dot{\tilde{g}}_t(z) = -\partial_t(\bar{g}_t(-\bar{z})) = -\overline{\partial_t g_t(-\bar{z})} = -\frac{2}{\bar{g}_t(-\bar{z}) - u(t)}$$

$$= \frac{2}{-\bar{g}_t(-\bar{z}) + u(t)} = \frac{2}{\tilde{g}_t(z) + u(t)}.$$

This is the chordal Loewner evolution with the driving function $\tilde{u}(t) = -u(t)$. □

Solution of 5.8. Since $f_t(z) = e^{-t}z + \cdots$,

$$\tilde{f}_t(z) = (e^{-tn}z^n + \cdots)^{1/n} = e^{-t}z + \cdots,$$

which means that $\tilde{f}_t$ has the right normalization. Since $f_t(\mathbb{D}) \subset f_s(\mathbb{D})$ for $s \le t$, we have

$$\tilde{f}_t(\mathbb{D}) = (f_{tn}(\mathbb{D}))^{1/n} \subset (f_{sn}(\mathbb{D}))^{1/n} = \tilde{f}_s(\mathbb{D}),$$

where we use the n-valued root. This proves that $\tilde{f}_t$ forms a Loewner chain.

Differentiating $\tilde{f}_t$, we get

$$\begin{aligned}
\partial_t \tilde{f}_t(z) &= (f_{nt}(z^n))^{(1-n)/n}\, \dot{f}_{nt}(z^n) \\
&= -z^n\,(f_{nt}(z^n))^{(1-n)/n}\, f'_{nt}(z^n)p_{nt}(z^n) \\
&= -z^n p_{nt}(z^n) z^{1-n}((f_{nt}(z^n))^{(1-n)/n}\, f'_{nt}(z^n)z^{n-1}) \\
&= -z p_{nt}(z^n)\tilde{f}'_t(z).
\end{aligned}$$

This proves that $\tilde{p}_t(z) = p_{nt}(z^n)$. □

Solution of 5.9. Let us consider the equation for g_t. The integral term is of the form

$$\int \frac{e^{i\theta} + w}{e^{i\theta} - w}\frac{d\theta}{2\pi},$$

where $w = g_t(z)$. This integral is identically equal to 1. This can be seen either from residue computation or from the fact that by the Schwarz formula this integral represents an analytic function such that its real part is equal to 1 on the boundary and at the origin. This simplifies the equation to

$$\dot{g}_t(z) = m_t g_t(z),$$

which trivially integrates to

$$g_t(z) = g_0(z)\exp\Big(\int_0^t m_t \mathrm{d}t\Big) = z\exp\Big(\int_0^t m_t \mathrm{d}t\Big).$$

This implies that $\Omega_t = \exp(-\int_0^t m_t \mathrm{d}t)\mathbb{D}$. In particular, when $m_t = 1$, it becomes $g_t(z) = e^t z$. □

Solution of 5.10. Let us consider the function $\tilde{g}_t(z) = g_t(z - x_0) + x_0$. It was established in Exercise 3.16 that this map corresponds to hulls $\tilde{K}_t = K_t + x_0$. Differentiating $\tilde{g}_t$ with respect to t, we have

$$\begin{aligned}\partial_t \tilde{g}_t(z) &= \int \frac{\mu_t(\mathrm{d}x)}{g_t(z-x_0) - x} = \int \frac{\mu_t(\mathrm{d}x)}{\tilde{g}_t(z) - (x + x_0)} \\ &= \int \frac{\mu_t(\mathrm{d}x - x_0)}{\tilde{g}_t(z) - x} = \int \frac{\tilde{\mu}_t(\mathrm{d}x)}{\tilde{g}_t(z) - x}.\end{aligned}$$

This proves that $\tilde{g}_t$ is the solution of the Loewner equation driven by $\tilde{\mu}_t$. □

Solution of 5.11. The main idea is quite simple: Under the Loewner evolution, all points go down until they hit the real line. The fastest decay happens when the driving measure is concentrated right below the point, which means that the fastest decay corresponds to the vertical slit.

To simplify notation, we denote $\operatorname{Im} g_t(iy_0)$ by y_t. By considering the imaginary part of the Loewner evolution, we have

$$\dot{y}_t = -\int \frac{y_t \mu_t(\mathrm{d}x)}{|g_t(iy_0) - x|^2}.$$

This implies

$$\dot{y}_t \geq -\int \frac{\mu_t(\mathrm{d}x)}{y_t} = -\frac{2}{y_t}.$$

This differential inequality is easy to integrate:

$$y_t \geq y_0 - 2\sqrt{t}.$$

Since $\sup\{t : y_t > 0\} = T(iy_0) \leq t$, we must have $y_0 \leq 2\sqrt{t}$. □

Solution of 5.12. Let us assume that $\operatorname{hcap}(K) < y_0^2/2 - \varepsilon$ for some positive ε. There are slit hulls K_n such that $K_n \to K$ in the Carathéodory sense (see Remark 3.20).

We claim that for every $\delta > 0$, K_n contains points with imaginary parts greater than $y_0 - \delta$ for all sufficiently large n. Indeed, if this is not the case, then the δ-neighbourhood of iy_0 is not separated from infinity by infinitely many of K_n, and hence iy_0 is in the kernel of some subsequence. Since $\mathbb{H}\setminus K$ is the kernel of every subsequence of $\mathbb{H}\setminus K_n$, this would imply that $iy_0 \notin K$.

Summing up, we get that for arbitrarily small ε and δ, there is a slit hull $\tilde{K}$ such that $\mathrm{hcap}(\tilde{K}) < \mathrm{hcap}(K) + \varepsilon/2 < y_0^2/2 - \varepsilon/2$ and it contains a point with an imaginary part greater than $y_0 - \delta$. Since every slit is generated by a Loewner chain, by Exercise 5.11, the last condition implies $\mathrm{hcap}(\tilde{K}) \geq (y_0 - \delta)^2/2$. Combining all inequalities, we have $(y_0 - \delta)^2/2 < y_0^2/2 - \varepsilon/2$. For fixed y_0 and ε, we can always choose δ small enough so that this inequality is wrong. This contradiction implies that $\mathrm{hcap}(K) \geq y_0^2/2$. □

Solution of 5.13. We already know from Exercise 5.11 that $\mathrm{Im}\, z \leq 2\sqrt{t}$ for all $z \in K_t$. It remains to be shown that $\mathrm{Re}\, z \in [a, b]$. To show this, let us consider the real part of the Loewner evolution. We get

$$\dot{x}_t = \int_a^b \frac{(x_t - x)\mu_t(\mathrm{d}x)}{|g_t(z) - x|^2},$$

where $x_t = \mathrm{Re}\, g_t(z)$. If $x_0 > b$, then $\dot{x}_t > 0$ and $x_t > x_0$ for all $t < T(z)$. In this case, we can also estimate the derivative of $y_t = \mathrm{Im}\, g_t(z)$:

$$\dot{y}_t = -\int \frac{y_t \mu_t(\mathrm{d}x)}{|g_t(z) - x|^2} \geq -2y_t(x - b)^{-2}.$$

Integrating this inequality, we have

$$y_t \geq y_0 e^{-2t(x-b)^{-2}}.$$

This proves that $y_t > 0$ for all t; hence $T(z) = \infty$. This proves that points with $\mathrm{Re}\, z > b$ cannot be in K_t. The same argument works for z with $\mathrm{Re}\, z < a$. This completes the proof that $K_t \subset [a, b] \times [0, 2\sqrt{t}]$. □

Bibliography

Ahlfors, L. V. (1978). *Complex Analysis: An Introduction to the Theory of Analytic Functions of One Complex Variable*, 3rd edn., International Series in Pure and Applied Mathematics (McGraw-Hill Book Co., New York).

Ahlfors, L. V. (2010). *Conformal Invariants: Topics in Geometric Function Theory*, Reprint of the 1973 original, with a foreword by Peter Duren, F. W. Gehring and Brad Osgood (AMS Chelsea Publishing, Providence, RI).

Baernstein, A. (1974). Integral means, univalent functions and circular symmetrization, *Acta Math.* **133**, 1, pp. 139–169.

Banjai, L. and Trefethen, L. N. (2003). A multipole method for Schwarz–Christoffel mapping of polygons with thousands of sides, *SIAM J. Sci. Comput.* **25**, 3, pp. 1042–1065.

Bazilevich, I. (1937). Supplement to the papers 'Zum Koeffizientenproblem der schlichten Funktionen' and 'Sur les théorèmes de Koebe–Bieberbach', *Mat. Sb.* **2**, 44, pp. 689–698.

Bazilevich, I. (1951). On distortion and coefficient theorems for univalent functions, *Mat. Sb.* **28**, pp. 147–164.

Beliaev, D. (2008). Integral means spectrum of random conformal snowflakes, *Nonlinearity* **21**, 7, p. 1435.

Beliaev, D. and Smirnov, S. (2010). Random conformal snowflakes, *Ann. of Math.* **172**, 1, pp. 597–615.

Beurling, A. (1989). *The Collected Works of Arne Beurling, Vol. 1: Complex Analysis*, Contemporary Mathematicians, edited by L. Carleson, P. Malliavin, J. Neuberger and J. Wermer (Birkhäuser Boston, Inc., Boston, MA).

Bieberbach, L. (1916). Über die Koeffizienten derjenigen Potenzreihen, welche eine schlichte Abbildung des Einheitskreises vermitteln, *Berl. Ber.* **1916**, pp. 940–955.

Carathéodory, C. (1912). Untersuchungen über die konformen Abbildungen von festen und veränderlichen Gebieten, *Math. Ann.* **72**, 1, pp. 107–144.

Carathéodory, C. (1913). Über die Begrenzung einfach zusammenhängender Gebiete, *Math. Ann.* **73**, 3, pp. 323–370.

Carleson, L. and Jones, P. W. (1992). On coefficient problems for univalent functions and conformal dimension, *Duke Math. J.* **66**, 2, pp. 169–206.

Charzynski, Z. and Schiffer, M. (1960). A new proof of the Bieberbach conjecture for the fourth coefficient, *Arch. Rational Mech. Anal.* **5**, 1, pp. 187–193.

Chung, K. L. (2002). *Green, Brown, and Probability & Brownian Motion on the Line* (World Scientific Publishing Co., Inc., River Edge, NJ).

Clunie, J. and Pommerenke, C. (1967). On the coefficients of univalent functions, *Michigan Math. J.* **14**, pp. 71–78.

Conway, J. B. (1995). *Functions of One Complex Variable. II*, Graduate Texts in Mathematics, Vol. 159 (Springer-Verlag, New York).

de Branges, L. (1985). A proof of the Bieberbach conjecture, *Acta Math.* **154**, 1–2, pp. 137–152.

Doob, J. L. (2012). *Classical Potential Theory and Its Probabilistic Counterpart: Advanced Problems*, Grundlehren der mathematischen Wissenschaften, Vol. 262 (Springer Science & Business Media).

Drasin, D., Duren, P., and Marden, A. (eds.) (1986). *The Bieberbach Conjecture*, Mathematical Surveys and Monographs, Vol. 21 (American Mathematical Society, Providence, RI).

Driscoll, T. A. and Trefethen, L. N. (2002). *Schwarz–Christoffel Mapping*, Cambridge Monographs on Applied and Computational Mathematics, Vol. 8 (Cambridge University Press, Cambridge).

Duren, P. (1983). *Univalent Functions*, Grundlehren der mathematischen Wissenschaften, Vol. 259 (Springer Science & Business Media).

Fomenko, O. M. and Kuz'mina, G. V. (1986). The last 100 days of the Bieberbach conjecture, *Math. Intelligencer* **8**, 1, pp. 40–47.

Gamelin, T. W. (2001). *Complex Analysis*, Undergraduate Texts in Mathematics (Springer-Verlag, New York).

Garabedian, P. R. and Schiffer, M. (1955a). A coefficient inequality for schlicht functions, *Ann. of Math.* (2) **61**, pp. 116–136.

Garabedian, P. R. and Schiffer, M. (1955b). A proof of the Bieberbach conjecture for the fourth coefficient, *J. Rational Mech. Anal.* **4**, pp. 427–465.

Garnett, J. B. (1986). *Applications of Harmonic Measure*, University of Arkansas Lecture Notes in the Mathematical Sciences, Vol. 8 (John Wiley & Sons, Inc., New York).

Garnett, J. B. and Marshall, D. E. (2008). *Harmonic Measure*, New Mathematical Monographs, Vol. 2 (Cambridge University Press, Cambridge).

Goluzin, G. (1936). On distortion theorems in the theory of conformal mappings, *Mat. Sb.* **18**, 60, pp. 379–390.

Goluzin, G. M. (1969). *Geometric Theory of Functions of a Complex Variable*, Translations of Mathematical Monographs, Vol. 26 (American Mathematical Society, Providence, RI).

Gray, J. (1994). On the history of the Riemann mapping theorem, *Rend. Circ. Mat. Palermo* (2) *Suppl.* **34**, pp. 47–94.

Harnack, A. (1887). *Die Grundlagen der Theorie des Logarithmischen Potentiales und der Eindeutigen Potentialfunktion in der Ebene* (VG Teubner).

Hayman, W. K. (1955). The asymptotic behaviour of p-valent functions, *Proc. London Math. Soc.* (3) **5**, pp. 257–284.

He, Z.-X. and Schramm, O. (1993). Fixed points, Koebe uniformization and circle packings, *Ann. of Math.* (2) **137**, 2, pp. 369–406.

Hedenmalm, H. and Shimorin, S. (2005). Weighted Bergman spaces and the integral means spectrum of conformal mappings, *Duke Math. J.* **127**, 2, pp. 341–393.

Hocking, J. G. and Young, G. S. (1988). *Topology*, 2nd edn. (Dover Publications, Inc., New York).

Kager, W., Nienhuis, B., and Kadanoff, L. P. (2004). Exact solutions for Loewner evolutions, *J. Statist. Phys.* **115**, 3–4, pp. 805–822.

Kemppainen, A. (2017). *Schramm–Loewner Evolution*, Springer, pp. 69–100.

Koebe, P. (1912). Ueber eine neue Methode der konformen Abbildung und Uniformisierung, *Nachr. Ges. Wiss. Göttingen, Math.-Physik. Kl.* **1912**, pp. 844–848.

Koralov, L. B. and Sinai, Y. G. (2007). *Theory of Probability and Random Processes*, 2nd edn., Universitext (Springer, Berlin).

Kufarev, P. P. (1947). A remark on integrals of Löwner's equation, *Doklady Akad. Nauk SSSR* (N.S.) **57**, pp. 655–656.

Lalley, S., Lawler, G., and Narayanan, H. (2009). Geometric interpretation of half-plane capacity, *Electron. Commun. Probab.* **14**, pp. 566–571.

Lawler, G. F. (2005). *Conformally Invariant Processes in the Plane*, Mathematical Surveys and Monographs, Vol. 114 (American Mathematical Society, Providence, RI).

Lind, J. R. (2005). A sharp condition for the Loewner equation to generate slits, *Ann. Acad. Sci. Fenn. Math.* **30**, 1, pp. 143–158.

Lind, J., Marshall, D. E. and Rohde, S. (2010). Collisions and spirals of Loewner traces, *Duke Math. J.* **154**, 3, pp. 527–573.

Löwner, K. (1923). Untersuchungen uber schlichte konforme Abbildungen des Einheitskreises. I, *Math. Ann.* **89**, pp. 103–121.

Makarov, N. G. (1998). Fine structure of harmonic measure, *Algebra i Analiz* **10**, 2, pp. 1–62.

Marshall, D. E. and Rohde, S. (2005). The Loewner differential equation and slit mappings, *J. Amer. Math. Soc.* **18**, 4, pp. 763–778.

Milin, I. M. (1977). *Univalent Functions and Orthonormal Systems*, Translations of Mathematical Monographs, Vol. 49 (American Mathematical Society, Providence, RI), translated from the Russian.

Natanson, I. P. (2016). *Theory of Functions of a Real Variable* (Courier Dover Publications).

Nehari, Z. (1975). *Conformal Mapping* (Dover Publications, Inc., New York).

Ohtsuka, M. (1970). *Dirichlet Problem, Extremal Length, and Prime Ends*, Van Nostrand Reinhold Mathematical Studies, Vol. 22. (Van Nostrand Reinhold Company, New York).

Osgood, W. (1900). On the existence of the Green's function for the most general simply connected plane region, *Trans. Amer. Math. Soc.* **1**, 3, pp. 310–314.

Ozawa, M. (1969). An elementary proof of the Bieberbach conjecture for the sixth coefficient, *Kōdai Math. Sem. Rep.* **21**, pp. 129–132.

Pederson, R. N. (1968/1969). A proof of the Bieberbach conjecture for the sixth coefficient, *Arch. Rational Mech. Anal.* **31**, pp. 331–351.

Pederson, R. and Schiffer, M. (1972). A proof of the Bieberbach conjecture for the fifth coefficient, *Arch. Rational Mech. Anal.* **45**, pp. 161–193.

Pommerenke, C. (1966). On the Loewner differential equation, *Michigan Math. J.* **13**, pp. 435–443.

Pommerenke, C. (1967). On the coefficients of univalent functions, *J. London Math. Soc.* **42**, pp. 471–474.

Pommerenke, C. (1975). *Univalent Functions: With a Chapter on Quadratic Differentials by Gerd Jensen*, Studia Mathematica: Mathematische Lehrbücher, Vol. 25 (Vandenhoeck & Ruprecht, Göttingen).

Radó, T. (1923). Über die Fundamentalabbildungen schlichter Gebiete. *Acta Litt. Sci. Szeged* **1**, pp. 240–251.

Riemann, B. (1851). *Grundlagen für eine Allgemeine Theorie der Functionen einer Veränderlichen Complexen Grösse*, Ph.D. thesis, Göttingen.

Riemann, B. (2004). *Collected Papers* (Kendrick Press, Heber City, UT), translated from the 1892 German edition by Roger Baker, Charles Christenson and Henry Orde.

Robertson, M. S. (1936). A remark on the odd schlicht functions, *Bull. Amer. Math. Soc.* **42**, 6, pp. 366–370.

Rohde, S. and Wong, C. (2014). Half-plane capacity and conformal radius, *Proc. Amer. Math. Soc.* **142**, 3, pp. 931–938.

Rudin, W. (1987). *Real and complex analysis*, 3rd edn. (McGraw-Hill Book Co., New York).

Schiffer, M. (1938). Sur un probleme d'extremum de la representation conforme, *Bull. Soc. Math. France* **66**, pp. 48–55.

Schramm, O. (2000). Scaling limits of loop-erased random walks and uniform spanning trees, *Israel J. Math.* **118**, 1, pp. 221–288.

Teichmüller, O. (1938). Untersuchungen über konforme und quasikonforme Abbildung, *Deutsche Math.* **3**, pp. 621–678.

Walsh, J. L. (1973). History of the Riemann mapping theorem, *Amer. Math. Monthly* **80**, pp. 270–276.

Weinstein, L. (1991). The Bieberbach conjecture, *Int. Math. Res. Notices* **1991**, 5, pp. 61–64.

Index

A

accessible point, 34
admissible metric, 130
argument principle, 4, 42, 56, 61
Arzelà–Ascoli theorem, 17

B

Beurling estimate, 153, 156, 206
Bieberbach conjecture, 112, 184
Bieberbach inequality, 89
Bieberbach theorem, 86
boundary correspondence, 33, 42
  analytic boundary, 41

C

Carathéodory boundary correspondence theorem, 41
Carathéodory class, 102, 190
Carathéodory convergence, 94, 212
Carathéodory convergence theorem, 95, 106
Carathéodory prime end theorem, 45
Carleson–Jones conjecture, 119
Christoffel–Schwarz formula, 29, 125, 164
  for rectangle, 31
  for triangle, 30
circular slit domain, 71
class Σ, 83
  coefficient estimates, 116
class S, 83
  compactness, 93
  coefficient estimates, 112, 183
complex sphere, *see* Riemann sphere, 6
conformal annulus, 64, 138
conformal automorphism, 15
conformal invariant, 66
conformal modulus, 66, 129, 131
  of doubly connected domain, 139
  reduced, 139
conformal radius, 23, 88, 139
conformal rectangle, 129, 137
cross-cut, 43, 157, 206
cross-ratio, 9

D

Dirichlet integral, 12
Dirichlet boundary problem, 48
Distortion theorem, 90
driving function, 174
driving measure, 190

E

$1/\theta$ estimate, 150
extremal annuli, 145
extremal distance, 137
  conjugated, 137
extremal length
  comparison rule, 134
  parallel rule, 135

serial rule, 134
symmetry rule, 136
extremal metric, 131, 133

F

fundamental sequence, 157

G

Grötzsch annulus, 145
Green's formula, 2, 70
Green's function, 54, 118, 145
Gronwall area theorem, 85
Growth theorem, 91

H

half-plane capacity, 26, 105, 181
integral representation, 106
of circular slit, 108
of half-disc, 105
of tilted slit, 107
of vertical slit, 105
scaling property, 105, 182
half-plane hull, 26, 105
harmonic measure, 57, 148
asymptotic behaviour, 150
lower bound, 155
upper bound, 151
Herglotz theorem, 103, 190
hull generated by curve, 209
Hurwitz theorem, 19, 21

I

inradius, 88
integral mean, 113, 117

J

Joukowsky function, 85

K

kernel convergence, *see* Carathéodory convergence, 94
kernel of a domain family, 94
Koebe 1/4 theorem, 87
Koebe distortion, 92
Koebe distortion theorem, 88
Koebe function, 84, 87, 89, 91
Koebe lemma, 36
Koebe root transform, 85, 87
Koebe transformation, 85, 89

L

length–area argument, 37, 45, 208
Lindelöf principle, 23
Littlewood theorem, 112
local growth property, 182
locally connected boundary, 41
Loewner chain, 189
Loewner differential equation, *see* Loewner evolution, 173
Loewner evolution
chordal, 181, 199
driven by measure, 190, 195, 199
driving function, 174, 181
external, 177
generated by curve, 209, 211
radial, 174, 195
scaling property, 179
whole-plane, 177
Loewner–Kufarev equation, *see* Loewner evolution, driven by measure, 190
logarithmic capacity, 104

M

Möbius transform, 85
Möbius transformation, 8, 14
mapping-out function, 26, 108
maximum modulus principle, 37
maximum principle, 3, 58, 61, 149, 156
metric, 130
Milin conjecture, 116
Montel theorem, 17, 21, 93

N

normal family, 16, 96, 172
null chain, 43, 159

P

parallel slit domain, 67
parametrization by capacity, 181
Poisson kernel, 49, 53
prime end, 43, 157–158
    impression, 44
propagation of convergence, 19

R

radial slit domain, 71
reduced modulus, 139
Riemann mapping theorem, 20
    doubly connected domains, 64
    multiply connected domains, 67
Riemann sphere, 6
Robertson conjecture, 115
Rotation theorem, 91
Rouché theorem, 4, 20, 96

S

schlicht function, *see* univalent function, 9, 83
Schwarz formula, 103, 107
Schwarz kernel, 49, 51, 53
Schwarz lemma, 14, 29, 88, 96
Schwarz reflection, 3, 26, 37, 108, 172
slit domain, 98, 170
spherical metric, 7
subordination, 15, 101, 189

T

Teichmüller annulus, 146
thermodynamic normalization, 25, 106, 201
trace, 209
transition function, 171, 191

U

univalent function, 9

V

Vitali theorem, 18

W

Whitney decomposition, 111
winding number, 4
Wolff's lemma, 37

Z

zero chain, *see* null chain, 43

www.ingramcontent.com/pod-product-compliance
Ingram Content Group UK Ltd.
Pitfield, Milton Keynes, MK11 3LW, UK
UKHW022333150426
470037UK00010B/71

9 781800 616356